Let's Review Regents:

Algebra II 2020

Gary M. Rubinstein
B.A. Mathematics
Tufts University
M.S. Computer Science
University of Colorado

About the Author

Gary Rubinstein has been teaching math for more than 25 years. He is a three-time recipient of the Math for America Master Teacher Fellowship. Gary lives with his wife Erica and his two children, Sarah and Sam. He has a YouTube channel at nymathteacher where students can find extra test tips and strategies for learning Algebra II.

Dedication
To Erica, Sarah, and Sam

Published by Kaplan, Inc., d/b/a Barron's Educational Series
750 Third Avenue
New York, NY 10017
www.barronseduc.com

ISBN: 978-1-5062-5387-9

Printed in Canada

10 9 8 7 6 5 4 3 2 1

Kaplan, Inc., d/b/a Barron's Educational Series print books are available at special quantity discounts to use for sales promotions, employee premiums, or educational purposes. For more information or to purchase books, please call the Simon & Schuster special sales department at 866-506-1949.

TABLE OF CONTENTS

PREFACE

In 2009, New York State adopted the Common Core Standards in order to qualify for President Obama's Race To The Top initiative. The Common Core math curriculum is more difficult than the previous math curriculum. The new state tests, including the Common Core Algebra II Regents, are more difficult as well.

Certain topics that had been in the Algebra II/Trigonometry curriculum for decades have been removed for not being rigorous enough. Other topics have been added with the goal of making 21st-century American students more career and college ready than their predecessors.

The main topics that have been cut from the curriculum are permutations, combinations, Bernoulli trials, binomial expansion, and the majority of the trigonometric identities like the sine sum, cosine sum, sine difference, cosine difference, and the different double-angle and half-angle formulas. Other topics have been moved into earlier grades like the law of sines and the law of cosines. More than half of the trigonometry that had once been part of Algebra II is no longer part of the course.

Other topics have been added to fill the gaps left by those topics now considered obsolete. Primarily, these new topics are often taught in AP Statistics as part of inferential statistics.

Aside from the change in topics, there is a change in the style of questions. Students now need to think more deeply about the topics because questions are intentionally phrased in a less straightforward way than they had been in the past.

Conquering the Algebra II test, something that was never an easy feat beforehand, has gotten much more difficult and will require more test preparation than before. Getting this book is a great first step toward that goal. In addition to reviewing all of the topics that can appear on this test, this book includes nearly 1,000 practice questions of various difficulty levels. This book can serve as a review or even as a way to learn the material for the first time. Teachers can also use this book to guide their pacing. They can focus on the types of questions that are most likely to appear on the test and spend less time on complicated aspects of the Common Core curriculum that are unlikely to be on the test.

The Common Core is part of a grand plan that is intended to propel our country to the top of the international rankings in math and reading. Good luck. We are all counting on you!

Gary Rubinstein
Math Teacher
2016

POLYNOMIAL EXPRESSIONS AND EQUATIONS

Chapter One

1.1 POLYNOMIAL ARITHMETIC

KEY IDEAS

A *polynomial* is an expression like $x^2 - 5x + 6$ that combines numbers and variables raised to different powers. Just as two numbers can be added, subtracted, multiplied, or divided, polynomials can be too.

Multiplying a Polynomial by a Constant

To multiply a polynomial by a constant, multiply the constant by each of the coefficients of the polynomial. This is sometimes called *distributing* the constant through the polynomial.

To multiply $3x^2 - 2x + 5$ by 4, multiply each of the coefficients by 4.

$$4 \cdot (3x^2 - 2x + 5) = 12x^2 - 8x + 20$$

Adding Polynomials

To add two polynomials, combine the *like terms*, which have the same variable raised to the same exponent.

To add the two polynomials $(x^2 - 5x + 6) + (2x^2 + 3x - 2)$, first combine the two x^2-terms, $x^2 + 2x^2 = 3x^2$. Then combine the two x-terms, $-5x + 3x = -2x$. Then combine the two constant terms $+6 - 2 = +4$. The sum is $3x^2 - 2x + 4$.

Subtracting Polynomials

Subtracting polynomials is more complicated than adding polynomials.

$$(3x^2 + 5x - 1) - (x^2 - 2x + 3)$$

Since the $-$ sign can be thought of as a negative 1 (-1) and the coefficient of the x^2 in the second polynomial is really a 1, this can be rewritten as:

$$(3x^2 + 5x - 1) - 1\,(1x^2 - 2x + 3)$$

Distribute the -1 through the second polynomial. The parentheses are no longer needed.

$$3x^2 + 5x - 1 - 1x^2 + 2x - 3$$

Combine like terms.

$$2x^2 + 7x - 4$$

Multiplying Polynomials

Multiplying two polynomials requires multiplying each combination of one term from the first polynomial with one term from the second polynomial and then combining all the products. The most common type of polynomial multiplication is when each of the polynomials has just two terms (called *binomials*). The four combinations can then be remembered with the word **FOIL**.

First terms
Outer terms
Inner terms
Last terms

$$(5x + 3)(2x - 4)$$

First terms in each expression: $5x \cdot 2x = +10x^2$

Outer terms in each expression: $5x \cdot -4 = -20x$

Inner terms in each expression: $+3 \cdot 2x = +6x$

Last terms in each expression: $+3 \cdot -4 = -12$

$$10x^2 - 20x + 6x - 12 = 10x^2 - 14x - 12$$

If one or both of the polynomials has more than two terms, then the **FOIL** method does not apply. Instead, get all the combinations by multiplying the first term in the polynomial on the left by all the terms in the polynomial on the right. Then multiply the second term in the polynomial on the left by all the terms in the polynomial on the right. Continue until the last term in the polynomial on the left has been multiplied by all the terms in the polynomial on the right.

Example 1

Multiply $(x + 3)(2x^2 - 4x + 5)$

Solution: First multiply the x in the binomial by each of the terms of $2x^2 - 4x + 5$. Then multiply the $+3$ in the binomial by each of the terms of $2x^2 - 4x + 5$. Combine the like terms.

$$(x \cdot 2x^2) + (x \cdot (-4x)) + (x \cdot 5) + (3 \cdot 2x^2) + (3 \cdot (-4x)) + (3 \cdot 5)$$

$$= 2x^3 - 4x^2 + 5x + 6x^2 - 12x + 15$$

$$= 2x^3 + 2x^2 - 7x + 15$$

2

Multiplication Patterns

Two useful patterns for multiplying binomials without using FOIL or the combination method are the perfect square multiplying pattern and the difference of perfect squares multiplying pattern.

Simplifying $(x + 5)^2$ with FOIL becomes $x^2 + 5x + 5x + (5 \cdot 5) = x^2 + 10x + 25$. The coefficient of the x in the solution is double the $+5$, whereas the constant in the solution is the square of $+5$. In general, $(x + a)^2 = x^2 + 2ax + a^2$.

Example 2

Use the perfect square multiplying pattern to simplify $(x - 3)^2$.

Solution: The coefficient will be $2 \cdot (-3) = -6$, and the constant will be $(-3)^2 = +9$.

$$x^2 - 6x + 9$$

When the only difference between two binomials is that one has a $+$ between the two terms and the other has a $-$ between the two terms, there is a shortcut for multiplying the binomials.

Simplifying $(x - 5)(x + 5)$ with FOIL becomes $x^2 + 5x - 5x - 25 = x^2 - 25$. There is no x-term in the answer, and the constant term is the negative square of the constant term of either of the binomials. In general, $(x - a)(x + a) = x^2 - a^2$.

Example 3

Use the difference of perfect squares multiplying pattern to simplify

$$(x - 3)(x + 3)$$

Solution: Since the only difference between the two binomials is the sign between the two terms, this pattern can be used. The answer is $x^2 - 3^2 = x^2 - 9$.

Dividing Polynomials

Dividing polynomials requires a process very similar to long division for numbers.

$$(2x^3 + x^2 - 11x + 12) \div (x + 3)$$

Step 1:

Set up for the long division process.

$$x + 3 \overline{)2x^3 + x^2 - 11x + 12}$$

Step 2:

Determine what you would need to multiply by the first term of the divisor (x in this example) to get the first term in the dividend ($2x^3$ in this example). Since you would need to multiply x by $2x^2$ to get $2x^3$, the first term of the solution is $2x^2$. Put that term over the x^2-term in the dividend.

$$x+3 \overline{\smash{\big)}\, 2x^3 + x^2 - 11x + 12} \overset{\displaystyle 2x^2}{}$$

Step 3:

Multiply the $2x^2$ by the $x + 3$ to get $2x^3 + 6x$. Put that product under the $2x^3 + x^2$. Subtract and bring down the $-11x$.

Step 4:

Determine what you would need to multiply by the first term of the divisor (x in this example) to get the first term in the expression at the bottom ($-5x^2$ in this example). Since you would need to multiply x by $-5x$ to get $-5x^2$, the second term of the solution is $-5x$. Put that term over the x-term in the dividend. Multiply the $-5x$ by the $x + 3$, and put the product under the $-5x^2 - 11x$. Then subtract and bring down the $+12$.

$$
\begin{array}{r}
2x^2 - 5x \\
x+3 \overline{\smash{\big)}\, 2x^3 + \ x^2 - 11x + 12} \\
-(2x^3 + 6x^2) \\
\hline
-5x^2 - 11x \\
-(-5x^2 - 15x) \\
\hline
4x + 12
\end{array}
$$

Step 5:

Determine what you would need to multiply by the first term of the divisor (x in this example) to get the first term in the expression at the bottom ($4x$ in this example). Since you would need to multiply x by 4 to get $4x$, the third term of the solution is +4. Put that term over the constant term in the dividend. Multiply the +4 by the $x + 3$, and put the product under the $4x + 2$. Then subtract. As there is nothing left to bring down, this final number at the bottom is the *remainder*. Since the remainder in this example is 0, we sometimes say there is no remainder and that $x + 3$ divides evenly into $2x^3 - x^2 - 11x + 12$.

$$
\begin{array}{r}
2x^2 - 5x + 4 \\
x+3{\overline{\smash{\big)}\,2x^3 + x^2 11x + 12}} \\
-(2x^3 + 6x^2) \\
\hline
-5x^2 - 11x \\
-(-5x^2 - 15x) \\
\hline
4x + 12 \\
-(4x + 12) \\
\hline
0
\end{array}
$$

Step 6:

You can check your answer by multiplying the solution by the divisor and then adding the remainder to see if the result is equal to the dividend.

$$(x + 3)(2x^2 - 5x + 4) + 0$$
$$= 2x^3 - 5x^2 + 4x + 6x^2 - 15x + 12 + 0$$
$$= 2x^3 + x^2 - 11x + 12$$

Example 4

What is the quotient and remainder (if any) of the following?

$$(3x^3 + 8x^2 - 14x + 13) \div (x + 4)$$

(1) $3x^2 + 4x - 2$, remainder 5
(2) $3x^2 - 4x + 2$, remainder 5
(3) $3x^2 + 4x + 2$, remainder 6
(4) $3x^2 - 4x - 2$, remainder 6

Solution:

$$
\begin{array}{r}
3x^2 - 4x + 2 \\
x+4{\overline{\smash{\big)}\,3x^3 + 8x^2 - 14x + 13}} \\
-(3x^3 + 12x^2) \\
\hline
-4x^2 - 14x \\
-(-4x^2 - 16x) \\
\hline
2x + 13 \\
-(2x + 8) \\
\hline
5
\end{array}
$$

The answer is choice (2).

Since Example 4 is a multiple-choice question, an alternative way to do this one would be to check each of the answer choices. Multiply each potential solution by $x + 4$, and add the potential remainder. The answer choice that gives you $3x^3 + 8x^2 - 14x + 13$ is the correct one.

Checking choice (2) would look like:

$$(x + 4)(3x^2 - 4x + 2) + 5$$
$$= 3x^3 - 4x^2 + 2x + 12x^2 - 16x + 8 + 5$$
$$= 3x^3 + 8x^2 - 14x + 13$$

The other choices all give different results. Admittedly, performing four multiplications to check the four choices might take longer than dividing. If on the test you forget how to divide polynomials and there is a multiple-choice question involving polynomial division, then this method would be a way to get the correct answer.

Check Your Understanding of Section 1.1

A. *Multiple-Choice*

1. What is $5 \cdot (2x^2 + 7x - 3)$?
   (1) $10x^2 + 35x - 15$ (3) $2x^2 + 7x - 15$
   (2) $10x^2 + 7x - 3$ (4) $2x^2 + 35x - 15$

2. What is $(-4) \cdot (2x^2 + 7x - 3)$?
   (1) $-8x^2 - 28x - 12$ (3) $-8x^2 + 7x - 3$
   (2) $-8x^2 - 28x + 12$ (4) $2x^2 + 7x + 12$

3. What is $(3x^2 - 5x + 7) + (2x^2 + 3x - 4)$?
   (1) $5x^2 - 8x + 3$ (3) $5x^2 - 2x + 11$
   (2) $5x^4 - 2x + 3$ (4) $5x^2 - 2x + 3$

4. What is $(5x^2 - 3x + 8) - (2x^2 + 4x - 2)$?
   (1) $3x^2 - 7x + 10$ (3) $3x^2 - 7x + 6$
   (2) $3x^2 + x + 10$ (4) $3x^2 + x + 6$

5. What is $2 \cdot (3x^2 - 4x + 7) - 3(x^2 - 2x - 5)$?
   (1) $3x^2 - 14x - 1$ (3) $3x^2 - 2x - 1$
   (2) $3x^2 - 14x + 29$ (4) $3x^2 - 2x + 29$

6. What is $(5x - 2)(2x + 7)$?
   (1) $10x^2 - 14$ (3) $10x^2 - 4x - 14$
   (2) $10x^2 + 35x - 14$ (4) $10x^2 + 31x - 14$

7. What is $(4x^2 + 2)(3x - 1)$?
   (1) $12x^3 - 4x^2 + 6x - 2$ (3) $12x^2 + 4x^2 - 6x + 2$
   (2) $12x^3 - 2$ (4) $12x^2 - 4x^2 - 6x - 2$

8. What is $(x + 2)(3x^2 + 2x - 7)$?
   (1) $3x^3 + 2x - 14$ (3) $3x^3 + 8x^2 + 3x - 14$
   (2) $3x^3 + 8x^2 - 3x - 14$ (4) $3x^3 - 8x^2 - 3x - 14$

9. What is $(x - 7)^2$?
   (1) $x^2 - 14x + 49$ (3) $x^2 + 49$
   (2) $x^2 - 14x - 49$ (4) $x^2 - 49$

10. What is $(3x^2 + 5)^2$?
   (1) $9x^4 + 25$ (3) $9x^4 + 15x - 25$
   (2) $9x^4 + 15x^2 + 25$ (4) $9x^4 + 30x^2 + 25$

11. What is $(x^3 + x^2 - 18x - 3) \div (x - 4)$?
  (1) $x^2 + 5x + 2$ R3
  (2) $x^2 + 5x + 2$ R4
  (3) $x^2 + 5x + 2$ R5
  (4) $x^2 + 5x + 2$ R6

B. *Show how you arrived at your answers.*

1. Simplify $5 \cdot (2x^2 - 3x + 4)$.

2. Zahra calculated $(5x^2 + 7x + 10) - (2x^2 + 3x + 4)$ as $5x^2 + 7x + 10 - 2x^2 + 3x + 4$ and got $3x^2 + 10x + 14$. What mistake did Zahra make?

3. If $(x + 5)(x + a) = x^2 + 7x + 10$, what is the value of a?

4. If $(x + a)^2 = x^2 + 10x + 25$, what is the value of a?

5. Simplify $(x + 3)^3$.

1.2   POLYNOMIAL FACTORING

KEY IDEAS

Factoring an integer means finding two other integers (other than 1) whose product is equal to the original integer. For example, the integer 15 has the *factors* 3 and 5 since $3 \cdot 5 = 15$. Likewise, a polynomial like $x^2 - 5x + 6$ can be factored into $(x - 2)(x - 3)$ because $(x - 2)(x - 3) = x^2 - 5x + 6$. When a polynomial is factored, the factors can provide useful information about the polynomial that was not apparent in the nonfactored form.

Just like some numbers can't be factored (for example, 7 and other prime numbers) some polynomials cannot be factored either. When a polynomial can be factored, there are several different methods of obtaining the factorization, depending on the polynomial.

Greatest Common Factor Factoring

Greatest common factor factoring is the first type of factoring you should always try. If all the terms of a polynomial have a common factor, that common factor can be "factored out." Often the only common factor is the number 1, in which case this type of factoring is not useful.

An example of using this method is factoring the polynomial $6x^3 - 4x^2 + 8x$. Each term of the polynomial $6x^3 - 4x^2 + 8x$ has a factor of $2x$. When the $2x$ is factored out, it is generally put on the left side and the remaining factor is put on the right in parentheses.

First write the $2x$ on the left of the parentheses

$$2x(\qquad\qquad) = 6x^3 - 4x^2 + 8x$$

Now imagine distributing the $2x$ through as if you were going to multiply the $2x$ by what needs to go inside the parentheses. For each term in the expression on the right, think "What does $2x$ need to be multiplied by to become this?" For the first term, $6x^3$, you have to multiply $2x$ by $3x^2$ ($2 \cdot 3 = 6$, and $x \cdot x^2 = x^3$). So the first term inside the parentheses is $3x^2$.

Do this for the other two terms to get

$$2x(3x^2 - 2x + 4) = 6x^3 - 4x^2 + 8x$$

Example 1

Factor the polynomial $8x^4 - 12x^3 + 16x^2$.

Solution: Each of the three terms has a factor of $4x^2$. The factorization is

$$4x^2(2x^2 - 3x + 4)$$

Factoring a Quadratic Trinomial into the Product of Two Binomials

What's the opposite of FOIL? Is it LIOF? No. The opposite is factoring a quadratic trinomial like $x^2 + 5x + 6$ into the product of two binomials.

If the trinomial is of the form $x^2 + bx + c = 0$, find two numbers that have a sum of b and a product of c. For $x^2 + 5x + 6$, the two numbers that have a sum of 5 and a product of 6 are $+2$ and $+3$. Those numbers become the constants of the two binomials $(x + 2)$ and $(x + 3)$. So $x^2 + 5x + 6 = (x + 2)(x + 3)$.

Example 2

Factor the polynomial $x^2 + 3x - 10$ into the product of two binomials.

Solution: The two numbers that have a sum of $+3$ and a product of -10 are -2 and $+5$. So the two factors are $(x - 2)$ and $(x + 5)$.

$$x^2 + 3x - 10 = (x - 2)(x + 5)$$

Perfect Square Trinomial Factoring

The trinomials $x^2 + 6x + 9$, $x^2 + 8x + 16$, and $x^2 + 10x + 25$ are three examples of perfect square trinomials. These can be factored into $(x + 3)^2$, $(x + 4)^2$, and $(x + 5)^2$, respectively. The way to recognize a perfect square trinomial of the form $x^2 + bx + c$ is to compare c to $\left(\dfrac{b}{2}\right)^2$:

If $c = \left(\dfrac{b}{2}\right)^2$, the trinomial is a perfect square and can be factored as $\left(x + \dfrac{b}{2}\right)^2$.

For $x^2 + 6x + 9$, since $9 = \left(\dfrac{6}{2}\right)^2$, the trinomial is a perfect square. It can be factored as $\left(x + \dfrac{6}{2}\right)^2 = (x+3)^2$.

Example 3

Which of the following is a perfect square trinomial?

    (1) $x^2 + 8x + 25$
    (2) $x^2 - 18x + 36$
    (3) $x^2 + 10x - 25$
    (4) $x^2 - 12x + 36$

Solution: Since $\left(-\dfrac{12}{2}\right)^2 = (-6)^2 = 36$, choice (4) is a perfect square trinomial. Notice that choice (3) would need to have +25 instead of −25 to be a perfect square trinomial.

Example 4

Factor $x^2 - 18x + 81$.

Solution: Since $\left(-\dfrac{18}{2}\right)^2 = (-9)^2 = 81$, the trinomial is a perfect square trinomial. It can be factored as $\left(x + \left(-\dfrac{18}{2}\right)\right)^2 = (x-9)^2$.

Difference of Perfect Squares Factoring

A quadratic expression like $x^2 - 5^2$ is known as the difference between perfect squares since each of the terms is a perfect square and there is a subtraction sign between the two signs. In general, the expression $x^2 - a^2$ can be factored into $(x - a)(x + a)$. For the example $x^2 - 5^2$, it factors into $(x - 5)(x + 5)$.

Example 5

Factor $x^2 - 81$.

   Solution: Since 81 can be written as 9^2, $x^2 - 81 = x^2 - 9^2 = (x - 9)(x + 9)$.

Factoring Cubic Expressions by Grouping

Polynomial expressions that have one of the variables raised to the third power are called cubic polynomials. Generally, they are very difficult to factor. Sometimes a technique called factor by grouping can be used to factor certain cubic polynomials.

   Factor by grouping is when you have a four-term cubic expression and you factor a common factor from the first two terms and another common factor from the last two terms. Then you cross your fingers and hope that there will be a new common factor that you can then factor out.

   For example, the polynomial $x^3 - 3x^2 + 4x - 12$ can be factored this way.

   An x^2 can be factored out of the first two terms, and a 4 can be factored out of the last two terms.

$$x^2(x - 3) + 4(x - 3)$$

   At this point, notice that both of the terms have a factor of $(x - 3)$. This can be factored out to become $(x - 3)(x^2 + 4)$ just like factoring an expression like $x^2 \cdot a + 4 \cdot a = a(x^2 + 4)$.

Example 6

Completely factor $x^3 + 2x^2 - 9x - 18$.

   Solution: First factor an x^2 out of the first two terms.

$$x^2(x + 2) - 9x - 18$$

   Now you can factor a -9 out of the second two terms so that the $(x + 2)$ factor matches the other one.

$$x^2(x + 2) - 9(x + 2)$$
$$(x + 2)(x^2 - 9)$$

Because the question says to factor "completely," check to see if any of the expressions can be factored further.

Since $x^2 - 9$ is a difference between perfect squares, it can be factored into $(x - 3)(x + 3)$. So the original expression can be completely factored to $(x + 2)(x - 3)(x + 3)$.

Factoring More Complicated Expressions

Polynomials that have exponents greater than 3 can sometimes be factored by rewriting the polynomials in an equivalent form that can be factored with other methods.

The expression $x^6 - 16$ has an exponent of 6. Since 6 is an even number, x^6 can be expressed as $(x^3)^2$. Since 16 is a perfect square, the original expression can be written as $(x^3)^2 - 4^2$, which can then be factored with the difference of perfect squares pattern: $(x^3 - 4)(x^3 + 4)$.

The trinomial $x^4 + 5x^2 - 14$ resembles the kind of quadratic trinomial from earlier in this chapter. It can be written as $(x^2)^2 + 5(x^2) - 14$. This looks a lot like $u^2 + 5u - 14$. If it were $u^2 + 5u - 14$, it would factor into $(u + 7)(u - 2)$. So the original trinomial factors instead into $(x^2 + 7)(x^2 - 2)$.

Example 7

Factor $x^8 + 10x^4 + 25$.

Solution: This can be expressed as $(x^4)^2 + 10(x^4) + 25$, which looks a lot like $u^2 + 10u + 25$. Since $u^2 + 10u + 25$ has the perfect square trinomial pattern, $25 = \left(\dfrac{10}{2}\right)^2$, it factors into $(u + 5)^2$. So the example factors into $(x^4 + 5)^2$.

Algebra Identities

When two algebraic expressions can be simplified to the same expression, it is called an identity. Proving that something is an identity requires simplifying one or both sides of the equation until the two sides are identical. Until an identity is proved, there will be a small question mark over the equal sign like $\overset{?}{=}$. After the identity is established, the question mark over the equals sign is replaced with a check $\overset{\checkmark}{=}$.

12

Example 8

Prove the following identity.

$$(x+y)^2 - 4xy \overset{?}{=} (x-y)^2$$

Solution:

$$(x+y)^2 - 4xy \overset{?}{=} (x-y)^2$$

$$x^2 + 2xy + y^2 - 4xy \overset{?}{=} x^2 - 2xy + y^2$$

$$x^2 - 2xy + y^2 \overset{\checkmark}{=} x^2 - 2xy + y^2$$

Number Theory Proofs

Sometimes a theorem about numbers can be proved by turning the theorem into an identity to be verified.

Example 9

The nth triangular number is $\dfrac{n(n+1)}{2}$. The nth square number is n^2.
Prove that the sum of the nth triangular number and the $(n+1)$st triangular number is equal to the $(n+1)$st square number.

Solution:

$$\frac{n(n+1)}{2} + \frac{(n+1)(n+2)}{2} \overset{?}{=} (n+1)^2$$

$$\frac{n^2 + n + n^2 + 3n + 2}{2} \overset{?}{=} (n+1)^2$$

$$\frac{2n^2 + 4n + 2}{2} \overset{?}{=} (n+1)^2$$

$$n^2 + 2n + 1 \overset{?}{=} (n+1)^2$$

$$(n+1)^2 \overset{\checkmark}{=} (n+1)^2$$

Check Your Understanding of Section 1.2

A. Multiple-Choice

1. Which shows $6x^3 - 15x^2 + 21x$ factored?
   (1) $3x(2x^2 - 5x + 7)$ (3) $2x(3x^2 - 15x^2 + 21x)$
   (2) $3x(2x^2 - 15x^2 + 21x)$ (4) $6x(x^2 - 2x^2 + 3x)$

2. Which shows $x^2 + 8x + 12$ factored?
   (1) $(x + 2)(x + 6)$ (3) $(x + 1)(x + 12)$
   (2) $(x + 3)(x + 4)$ (4) $(x + 3)(x + 5)$

3. Which shows $x^2 + 4x - 12$ factored?
   (1) $(x + 2)(x - 6)$ (3) $(x - 2)(x + 6)$
   (2) $(x - 3)(x + 4)$ (4) $(x - 1)(x - 12)$

4. Which shows $x^2 + 10x + 25$ factored?
   (1) $(x - 5)^2$ (3) $(x + 1)(x + 25)$
   (2) $(x + 5)(x - 5)$ (4) $(x + 5)^2$

5. Which shows $x^2 - 12x + 36$ factored?
   (1) $(x + 6)^2$ (3) $(x - 1)(x - 36)$
   (2) $(x + 6)(x - 6)$ (4) $(x - 6)^2$

6. Which shows $x^2 - 64$ factored?
   (1) $(x - 8)^2$ (3) $(x + 8)^2$
   (2) $(x - 8)(x + 8)$ (4) $(x - 8)(x - 8)$

7. Which shows $x^3 + 2x^2 + 6x + 12$ factored?
   (1) $(x + 6)(x^2 - 12)$ (3) $(x + 2)^2(x + 3)$
   (2) $(x + 2)(x^2 + 6)$ (4) $(x + 1)(x + 2)(x + 6)$

8. Which shows $x^3 - 5x^2 + 4x - 20$ factored?
   (1) $(x - 5)(x^2 + 4)$ (3) $(x - 2)^2(x + 5)$
   (2) $(x - 4)(x^2 + 5)$ (4) $(x - 1)(x + 4)(x + 5)$

9. Which shows $x^6 - 25$ factored?
   (1) $(x^3 - 5)(x^3 + 5)$ (3) $(x^3 - 5)^2$
   (2) $(x^6 - 5)(x^6 + 5)$ (4) $(x^3 + 5)^2$

10. Which shows $x^6 - 2x^3 - 24$ factored?
   (1) $(x^3 + 6)(x^3 - 4)$ (3) $(x^3 - 6)(x^3 + 4)$
   (2) $(x^3 - 8)(x^3 + 3)$ (4) $(x^3 + 8)(x^3 - 3)$

B. *Show how you arrived at your answers.*

1. Factor $2x^2 + 7x + 3$.

2. Factor $x^2 + (17 + 39)x + (17 \cdot 39)$.

3. What value for c can $x^2 - 18x + c$ be factored into $(x - 9)^2$?

4. How can the fact that $24^2 - 5^2 = 551$ be used to find the factors (not including 1 or 551) of 551?

5. Completely factor $x^4 - 13x + 36$.

1.3 THE REMAINDER THEOREM AND THE FACTOR THEOREM

KEY IDEAS

When something is a factor of a number, like 5 is a factor of 10, there will be no remainder when the number is divided by the factor. This is called *the factor theorem*. When something is not a factor of a number, like 3 is not a factor of 10, there will be some remainder when the number is divided by the factor. With polynomial division, there is a theorem called *the remainder theorem* that enables you to determine the remainder of some divisions without going through the long division process.

The Remainder Theorem

If you divide the polynomial function $f(x) = x^3 - 5x^2 + 6x - 3$ by $(x - 4)$ with the long division process, you get $x^2 - x + 2$ remainder 5. The remainder theorem says that you will also get the number 5 if you substitute $+4$ (the opposite of the -4 in the $(x - 4)$ divisor) into the $f(x) = x^3 - 5x^2 + 6x - 3$.

To check if the remainder theorem works for this example, evaluate $f(4) = 4^3 - (5 \cdot 4^2) + (6 \cdot 4) - 3 = 64 - 80 + 24 - 3 = 5$.

================= **MATH FACTS** =================

The remainder theorem says that the remainder when a polynomial equation is divided by $(x - a)$ is equivalent to the value of the polynomial when $+a$ is substituted for x. If the expression is $(x + a)$, then substitute $-a$ into the polynomial.

Example 1

What is the remainder when $f(x) = 4x^3 - 9x^2 + 7x + 1$ is divided by $(x + 5)$?

   (1) 5
   (2) −5
   (3) $f(5)$
   (4) $f(-5)$

Solution: The answer is choice (4). The remainder theorem states that the remainder will be the same as whatever the function evaluates to for the opposite of the constant in the divisor. For this example, $f(-5) = (4 \cdot (-5)^3) - (9 \cdot (-5)^2) + 7(-5) + 1 = -759$.

The Factor Theorem

When a binomial like $(x - 2)$ is a factor of a polynomial like $4x^3 - 3x^2 - 17x + 14$, it means that there will be a remainder of 0 when $4x^3 - 3x^2 - 17x + 14$ is divided by $x - 2$. If you evaluate the polynomial $4x^3 - 3x^2 - 17x + 14$ for $x = 2$, it becomes $(4 \cdot 2^3) - (3 \cdot 2^2) - (17 \cdot 2) + 14 = 32 - 12 - 34 + 14 = 0$, just as the remainder theorem predicted. From this, we get the factor theorem.

================= **MATH FACTS** =================

The factor theorem says that if $(x - a)$ is a factor of a polynomial, then the value of the polynomial when $+a$ is substituted for x will be 0.

Example 2

$(x - 3)$ is a factor of which polynomial?

   (1) $f(x) = x^3 - 9x^2 + 25x - 19$
   (2) $f(x) = x^3 - 9x^2 + 25x - 20$
   (3) $f(x) = x^3 - 9x^2 + 25x - 21$
   (4) $f(x) = x^3 - 9x^2 + 25x - 22$

Solution: Choice (3) is the answer. Evaluate each choice at $x = 3$. Choice (3) becomes $3^3 - (9 \cdot 3^2) + (25 \cdot 3) - 21 = 27 - 81 + 75 - 21 = 0$. The factor theorem says that if $f(3) = 0$, then $(x - 3)$ is a factor of $f(x)$.

Check Your Understanding of Section 1.3

A. *Multiple-Choice*

1. What is the remainder when $x^3 + x^2 - 9x - 5$ is divided by $x - 3$?
   (1) 4   (2) 5   (3) 6   (4) 7

2. What is the remainder when $2x^3 + 7x^2 - 10x + 28$ is divided by $x + 5$?
   (1) 2   (2) 3   (3) 4   (4) 5

3. If $f(x) = x^3 - 7x^2 + 5x - 9$, what is the remainder when $f(x)$ is divided by $(x + 4)$?
   (1) $f(4)$ (2) $f(-4)$ (3) $f(9)$ (4) $f(-9)$

4. If the remainder when $x^3 - 2x^2 + ax - 3$ is divided by $x - 2$ is 7, what is the value of a?
   (1) 2   (2) 3   (3) 4   (4) 5

5. If the remainder when $x^3 + ax^2 - 5x + 4$ is divided by $x - 3$ is 97, what is the value of a?
   (1) 7   (2) 8   (3) 9   (4) 10

6. If $(x + 4)(x - 5) = x^2 - x - 20$, what is the remainder when $x^2 - x - 20$ is divided by $(x + 4)$?
   (1) 3   (2) 2   (3) 1   (4) 0

7. Which of the following is a factor of $x^3 + 3x^2 - 10x - 24$?
   (1) $(x - 1)$ (2) $(x - 2)$ (3) $(x - 3)$ (4) $(x - 4)$

8. Which of the following is a factor of $2x^4 - 9x^3 - 9x^2 + 46x + 24$?
   (1) $(x - 3)$ (2) $(x + 4)$ (3) $(x - 2)$ (4) $(x - 1)$

9. If $(x - 5)$ is a factor of $x^3 - 4x^2 + 7x + a$, what is the value of a?
   (1) 20   (2) -20   (3) 60   (4) -60

10. If $(x + 2)$ is a factor of $x^3 + ax^2 - 5x - 26$, what is the value of a?
    (1) 4   (2) 5   (3) 6   (4) 7

B. *Show how you arrived at your answers.*

1. What is the remainder when $x^3 + 5x^2 - 7x + 3$ is divided by $x - 4$?

2. $f(x) = x^4 - 3x^3 + 6x - 11$. If $f(3) = 7$, what is the remainder when $x^4 - 3x^3 + 6x - 11$ is divided by $(x - 3)$?

3. If the remainder when $x^3 + 4x^2 - 9x + 2$ is divided by $(x - 4)$ is 94, what is the remainder be when $x^3 + 4x^2 - 9x + 8$ is divided by $x - 4$?

4. Zoe and Jose tried to figure out if $x - 7$ is a factor of $x^3 - 12x^2 + 43x - 56$. Zoe did it by dividing, and Jose did it more quickly with the remainder theorem. How did Jose do it?

5. If $f(x) = x^4 + x^3 - x^2 + 11x - 12$, then $f(-3) = 0$. What is one factor of $f(x)$?

1.4 POLYNOMIAL EQUATIONS

$$\wedge$$
KEY IDEAS

A *polynomial equation* like $x^2 - 5x + 6 = 0$ involves an equal sign with a polynomial expression on one or both sides. The solution set of a polynomial equation is the set of numbers that make the left side of the equation equal to the right side of the equation. Polynomial equations usually have more than one solution.

Solving Quadratic Equations that Have No x-Term

A *quadratic equation* is one where the largest exponent is a 2. The simplest type of quadratic equation is when there is no x-term, such as the quadratic equation $x^2 = 9$.

To solve $x^2 = 9$, take the square root of each side. Be sure to put a $\pm$ on the square root on the right-hand side of the equal sign to account for the fact that $(+3)^2$ and $(-3)^2$ both equal $+9$.

$$x^2 = 9$$
$$\sqrt{x^2} = \pm\sqrt{9}$$
$$x = \pm 3$$

The solution set is $\{3, -3\}$.

Solving Factored Polynomial Equations

The equation $(x - 2)(x + 3) = 0$ is a quadratic equation. If the left side was simplified, the highest exponent would be 2.

Equations like this with two or more factors on the left side of the equation and a zero on the right side can be solved very quickly.

The only way $(x - 2)(x + 3)$ can equal zero is if one of the two factors is also equal to zero.

So $(x - 2)(x + 3)$ will equal zero if $x - 2$ is 0 or if $x + 3 = 0$.

$$(x - 2)(x - 3) = 0$$

$$x - 2 = 0 \quad \text{or} \quad x - 3 = 0$$

$$\underline{+2 = +2} \qquad \underline{+3 = +3}$$

$$x = 2 \quad \text{or} \quad x = 3$$

The solution set is $\{2, 3\}$.

Example 1

What is the solution set to the equation $(x - 1)(x - 3) = 0$?

Solution:

$$x - 1 = 0 \quad \text{or} \quad x - 3 = 0$$
$$x = 1 \quad \text{or} \quad x = 3$$

The solution set is $\{1, 3\}$.

Example 2

What is the solution set to the equation $(x - 1)(x - 3)(x + 5) = 0$?

Solution: With three factors, it is still true that if any one of them is equal to zero, the product of the three factors is also equal to zero. So the solution set can be found by using the three equations.

$$x - 1 = 0 \qquad \text{or} \qquad x - 3 = 0 \qquad \text{or} \qquad x + 5 = 0$$
$$x = 1 \qquad \text{or} \qquad x = 3 \qquad \text{or} \qquad x = -5$$

The solution set is $\{1, 3, -5\}$.

Solving Quadratic Equations by Factoring

Not all quadratic polynomials factor. If one does in an equation where there is a zero on the right-hand side of the equal sign, the solution set can be found very quickly.

$$x^2 - 5x + 6 = 0$$

If possible, start by factoring the left-hand side:

$$(x - 2)(x - 3) = 0$$

$$x - 2 = 0 \quad \text{or} \quad x - 3 = 0$$

$$\underline{+2 = +2} \qquad \underline{+3 = +3}$$

$$x = 2 \quad \text{or} \quad x = 3$$

The solution set is {2, 3}.

Example 3

Find the solution set to the equation $x^2 - 2x - 3 = 5$.

Solution: Even though the left-hand side of this equation can factor, do not factor until the right-hand side of the equation is a zero. Do this by subtracting 5 from both sides of the equation.

$$x^2 - 2x - 3 = 5$$

$$\underline{-5 = -5}$$

$$x^2 - 2x - 8 = 0$$

$$(x - 4)(x + 2) = 0$$

$$x - 4 = 0 \quad \text{or} \quad x + 2 = 0$$

$$x = 4 \quad \text{or} \quad x = -2$$

The solution set is {4, −2}.

The two solutions, 4 and −2, are the opposites of the two constants in the factors, $(x - 4)$ and $(x + 2)$. In general, if the *factors* of a quadratic expression are $(x - a)$ and $(x - b)$, then the *zeros* or *roots* of the quadratic expression are $x = a$ and $x = b$. This also works in reverse. If the zeros of a quadratic expression are $x = a$ and $x = b$, then the factors of the expression are $(x - a)$ and $(x - b)$.

Example 4

Find a quadratic expression whose roots are +5 and −3.

Solution: If the roots are +5 and −3, the factors are $(x - 5)$ and $(x + 3)$. So the quadratic expression could be $(x - 5)(x + 3)$, which can be simplified to $x^2 - 2x - 15$.

Example 5

What are the roots of a cubic expression that has the factors $(x + 6)$, $(x - 3)$, and $(x - 5)$?

Solution: Set each of the factors equal to zero to find the roots.

$$(x + 6) = 0 \qquad (x - 3) = 0 \qquad (x - 5) = 0$$
$$x = -6 \qquad\quad x = 3 \qquad\quad x = 5$$

The roots are $\{-6, 3, 5\}$.

Solving Quadratic Equations with the Quadratic Formula

When the quadratic expression does not factor, the equation has irrational roots and can be solved with the quadratic formula.

================= MATH FACTS =================

The two solutions to the quadratic equation $ax^2 + bx + c = 0$ can be found with the quadratic formula: $x = \dfrac{-b \pm \sqrt{b^2 - 4ac}}{2a}$

*Note: Simplifying radicals is explained in more detail in Chapter 4.

An example of a quadratic equation where the quadratic expression cannot factor is $x^2 - 4x + 1 = 0$, where $a = 1$, $b = -4$, and $c = 1$.
According to the quadratic formula:

$$x = \frac{-(-4) \pm \sqrt{(-4)^2 - 4(1)(1)}}{2(1)} = \frac{4 \pm \sqrt{16 - 4}}{2} = \frac{4 \pm \sqrt{12}}{2} = \frac{4 \pm 2\sqrt{3}}{2} = 2 \pm \sqrt{3}$$

The solution set is $\{2 + \sqrt{3}, 2 - \sqrt{3}\}$.

Example 6

Use the quadratic formula to find the two solutions to the equation $x^2 - 6x + 7 = 0$.

Solution:

$$x = \frac{-(-6) \pm \sqrt{(-6)^2 - 4(1)(7)}}{2(1)} = \frac{6 \pm \sqrt{36 - 28}}{2} = \frac{6 \pm \sqrt{8}}{2} = \frac{6 \pm 2\sqrt{2}}{2} = 3 \pm \sqrt{2}$$

21

Check Your Understanding of Section 1.4

A. Multiple-Choice

1. What is the solution set of $x^2 = 16$?
   (1) $\{4\}$ (2) $\{4, -4\}$ (3) $\{8\}$ (4) $\{8, -8\}$

2. What is the solution set of $(x - 4)(x - 7) = 0$?
   (1) $\{4, 7\}$ (2) $\{-4, -7\}$ (3) $\{-4, 7\}$ (4) $\{4, -7\}$

3. What is the solution set of $(x - 3)(x + 9) = 0$?
   (1) $\{-3, 9\}$ (2) $\{3, 9\}$ (3) $\{-3, -9\}$ (4) $\{3, -9\}$

4. What is the solution set of $(x - 2)(x - 5)(x + 1) = 0$?
   (1) $x = -2, x = -5, x = 1$ (3) $x = -2, x = 5, x = -1$
   (2) $x = 2, x = -5, x = 1$ (4) $x = 2, x = 5, x = -1$

5. What is the solution set of $x^2 + 10x + 24 = 0$?
   (1) $x = -4, x = -6$ (3) $x = -4, x = 6$
   (2) $x = 4, x = 6$ (4) $x = 4, x = -6$

6. What is the solution set of $x^2 + 3x + 2 = 12$?
   (1) $x = -5, x = 2$ (3) $x = 5, x = -2$
   (2) $x = -1, x = -2$ (4) $x = 1, x = 2$

7. Which equation has the solutions $x = 5, x = -2$?
   (1) $(x + 5)(x - 2) = 0$ (3) $(x - 5)(x + 2) = 0$
   (2) $(x + 5)(x + 2) = 0$ (4) $(x - 5)(x - 2) = 0$

8. Which equation has the solutions $x = -4, x = 3$?
   (1) $x^2 - x - 12 = 0$ (3) $x^2 - x + 12 = 0$
   (2) $x^2 + x + 12 = 0$ (4) $x^2 + x - 12 = 0$

9. What are the solutions to $x^2 - 6x + 4 = 0$?
   (1) $x = 5 \pm \sqrt{3}$ (3) $x = 3 \pm \sqrt{5}$
   (2) $x = 3 \pm \sqrt{2}$ (4) $x = 2 \pm \sqrt{5}$

10. What are the solutions to $x^2 - 10x + 23$?
   (1) $x = 2 \pm \sqrt{5}$ (3) $x = 3 \pm \sqrt{2}$
   (2) $x = 5 \pm \sqrt{2}$ (4) $x = 5 \pm \sqrt{3}$

B. *Show how you arrived at your answers.*

1. Lila solves the equation $x^2 - 25 = 0$ by first factoring $x^2 - 25$ into $(x - 5)(x + 5)$. Skylar solves the same equation by first adding 25 to both sides of the equation to get $x^2 = 25$. Who is right?

2. What are the three solutions to $x^3 + 4x^2 - 9x - 36 = 0$?

3. Create an equation that has the three solutions $x = 2$, 4, and 5.

4. In terms of b, what is the solution to the quadratic equation $x^2 + bx - 4 = 0$?

5. Noelle used the quadratic formula to solve $x^2 + 10x + 16 = 0$. Delilah solved it without the quadratic formula. What did Delilah notice that enabled her to solve this equation without using the quadratic formula?

1.5 QUADRATIC GRAPHS

⌃
KEY IDEAS
⟋ ⟍

A graph of the solution set of a two-variable quadratic equation, like $y = x^2 - 5x + 6$, has a shape called a *parabola*. To produce an accurate sketch of this graph, determine the *vertex* of the parabola and the x-intercepts of the parabola if any exist. Depending on the form of the equation, you can use several methods to produce this graph.

The most basic quadratic graph is the graph of $y = x^2$. Some of the ordered pairs in the solution set of this equation are (0, 0), (1, 1), (2, 4), and (3, 9). Since the square of a negative number is positive, ordered pairs with negative x-coordinates will have positive y-coordinates. So (−1, 1), (−2, 4), and (−3, 9) are three more ordered pairs. When these seven ordered pairs are graphed, they can be connected with a U-shaped curve called a *parabola*. The lowest point, which is (0, 0) for this parabola, is the parabola's *vertex*.

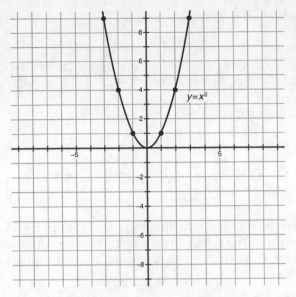

This graph is also the graph of the *quadratic function* $f(x) = x^2$.

Changing the *a* Coefficient

If a coefficient greater than 1 is in front of the x^2, such as $y = 2x^2$, the graph is still a parabola. However, it is narrower than the one created by $y = x^2$. The coefficient is usually called *a*, so the equation is $y = ax^2$. The vertex is still at $(0, 0)$.

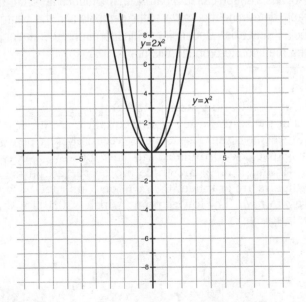

If *a* is negative, the parabola will be reflected over the *x*-axis and resemble an upside-down "U."

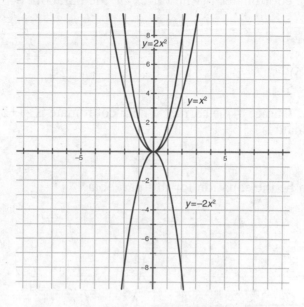

Vertex Form of a Quadratic Equation

A quadratic equation like $y = 1(x - 2)^2 - 5$ is said to be in *vertex form*. Vertex form is an expression like $y = a(x - h)^2 + k$, where *a*, *h*, and *k* are replaced with numbers. When a quadratic equation is in vertex form, the *x*-coordinate of the vertex is the opposite of the constant in the parentheses and the *y*-coordinate of the vertex is the constant outside the parentheses. For $y = 1(x - 2)^2 - 5$, the vertex is $(-(-2), -5) = (2, -5)$.

MATH FACTS

The vertex of the parabola defined by $y = a(x - h)^2 + k$ is (h, k). If *a* is greater than 1, the parabola will be narrow. If *a* is between 0 and 1, the parabola will be wide. If *a* is negative, the parabola will resemble an upside-down "U."

Example 1

What are the coordinates of the vertex of the parabola defined by $y = 3(x + 2)^2 - 5$?

   (1) (2, 5)
   (2) (2, −5)
   (3) (−2, 5)
   (4) (−2, −5)

Solution: Choice (4) is correct. The *x*-coordinate is −(+2) = −2, and the *y*-coordinate is −5.

Example 2

Which could be the equation for this parabola?

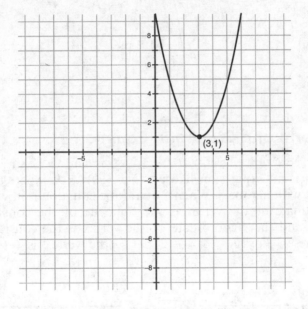

   (1) $y = 1 (x + 3)^2 + 1$
   (2) $y = 1 (x + 3)^2 - 1$
   (3) $y = 1 (x - 3)^2 + 1$
   (4) $y = 1 (x - 3)^2 - 1$

Solution: Choice (3) is correct. The constant inside the parentheses is the opposite of the *x*-coordinate of the vertex. The constant outside the parentheses is the *y*-coordinate.

Graphing Quadratic Equations

To sketch a more accurate graph, find the *x*-intercepts and *y*-intercept of the parabola. To find the *x*-intercepts, set $y = 0$ and solve for the values (if there are any) of x. The y-intercept can be found by setting $x = 0$ and solving for y.

Example 3

Sketch a graph, including the vertex and the x-intercepts (if any) and y-intercept, of $y = (x - 3)^2 - 1$.

Solution: The vertex is $(3, -1)$. To find the x-intercepts, solve the equation $0 = (x - 3)^2 - 1$:

$$\begin{aligned}
0 &= (x-3)^2 - 1 \\
\underline{+1 = \qquad\quad +1} \\
1 &= (x-3)^2 \\
\pm\sqrt{1} &= \sqrt{(x-3)^2} \\
\pm 1 &= x - 3 \\
\underline{+3 = \quad +3} \\
3 \pm 1 &= x \\
4 &= x \quad \text{or} \quad 2 = x
\end{aligned}$$

The two x-intercepts are $(4, 0)$ and $(2, 0)$.

The y-intercept can be found with the equation $y = (0 - 3)^2 - 1 = 9 - 1 = 8$. So the y-intercept is $(0, 8)$.

Graph the vertex, the two x-intercepts, and the y-intercept. Sketch a parabola that passes through the four points.

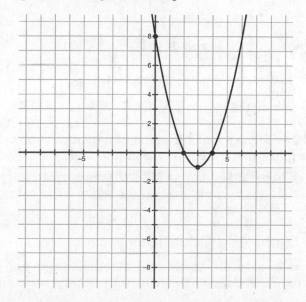

Vertex form is related to graphs of transformed functions. The quadratic function $g(x) = 1(x - 2)^2 - 5$ is related to the basic quadratic function $f(x) = x^2$. The $g(x)$ function can be expressed as $g(x) = f(x - 2) - 5$. The graph of a function defined this way is a translation of the original graph 2 units to the right and 5 units down. When the vertex of the graph of $f(x) = x^2$, which is $(0, 0)$, is translated 2 units to the right and 5 units down, the vertex of the transformed parabola will be $(2, -5)$.

Standard Form of a Quadratic Equation

Most quadratic functions and two-variable quadratic equations are not written in vertex form. Instead, they look like $f(x) = x^2 - 5x + 6$ or $y = x^2 - 5x + 6$. An equation in this simplified form is called *standard form*, $ax^2 + bx + c$. When the equation is in standard form, determining the vertex takes more work.

MATH FACTS

The x-coordinate of the vertex of the graph of $f(x) = ax^2 + bx + c$ is $\dfrac{-b}{2a}$. The

y-coordinate of the vertex is the value of $f\left(\dfrac{-b}{2a}\right)$. The vertex of the

parabola defined by $f(x) = x^2 - 6x + 8$, has an x-coordinate of

$\dfrac{-(-6)}{2 \cdot 1} = 3$ and a y-intercept of $f(3) = -1$.

Example 4

What is the vertex of the graph of $f(x) = 2x^2 - 12x + 3$?

Solution: The x-coordinate of the vertex is $\dfrac{-b}{2a} = \dfrac{-(-12)}{2 \cdot 2} = \dfrac{12}{4} = 3$.
The y-coordinate of the vertex is $f(3) = 2 \cdot 3^2 - 12 \cdot 3 + 3 = -15$. So the vertex is at $(3, -15)$.

Finding the x-intercepts of the graph of a quadratic function in standard form requires factoring. For the graph of $f(x) = x^2 - 6x + 8$, the x-intercepts can be found with the equation $0 = x^2 - 6x + 8$:

$$0 = x^2 - 6x + 8$$

$$0 = (x - 4)(x - 2)$$

$$x - 4 = 0 \text{ or } x - 2 = 0$$

$$x = 4 \text{ or } x = 2$$

So the x-intercepts are $(4, 0)$ and $(2, 0)$.

The y-intercept can be found by setting $x = 0$ and solving for y, $y = 0^2 - 6 \cdot 0 + 8 = 8$.

So the y-intercept is $(0, 8)$.

Since the vertex was determined to be $(3, -1)$ in a previous Math Fact, the graph looks like this:

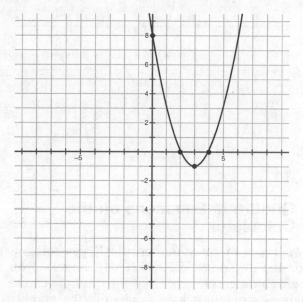

Example 5

Find the vertex and the x-intercepts of the graph of the function $f(x) = x^2 - 2x - 3$. Use these points to sketch the graph.

Solution: The x-coordinate of the vertex is $\dfrac{-(-2)}{2 \cdot 1} = +1$. The y-coordinate of the vertex is $f(1) = 1^2 - 2 \cdot 1 - 3 = -4$. The vetex is $(1, -4)$. The x-intercepts can be found with the equation $0 = x^2 - 2x - 3$:

$$0 = x^2 - 2x - 3$$

$$0 = (x + 1)(x - 3)$$

$$0 = x + 1 \text{ or } 0 = x - 3$$

$$-1 = x \text{ or } 3 = x$$

The x-intercepts are $(-1, 0)$ and $(3, 0)$.

The graph looks like the following:

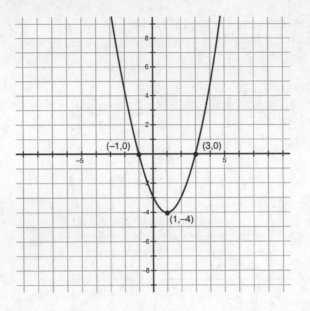

Graphing Quadratic Equations on the Graphing Calculator

There are two types of questions on the Regents Exam that involve graphing. For the questions in Parts II, III, and IV, you will have to produce the graph on graph paper. For the multiple-choice questions in Part I, though, you may just have to identify the graph from a list of choices. The graphing calculator can serve as a good way to check a graph that you produced by hand or to see what the graph looks like so you can select the correct graph from a set of choices.

For the TI-84:

Press the [Y=] button and input $x^2 - 5x + 6$ after the Y1=. Press [ZOOM] [6] to see the graph.

For the TI-Nspire:

From the home screen, press [B] to open the graphing Scratchpad. If the Entry Line is not visible, press [tab]. Input $x^2 - 5x + 6$ after f1(x)= and press [enter].

Focus and Directrix of a Parabola

Every parabola has an invisible line and invisible point associated with it. The invisible line is called the *directrix*, and the invisible point is called the *focus*. For a parabola that resembles a right-side-up "U," the focus is a bit above the vertex. The directrix is a horizontal line below the vertex. The distance between the focus and vertex is the same as the distance between the vertex and the directrix. For a parabola that resembles an upside-down "U," the focus is a bit below the vertex and the directrix is a bit above. Again, the distance between the focus and vertex is the same as the distance between the vertex and the directrix.

MATH FACTS

The distance between the vertex and the focus is the same as the distance between the vertex and the directrix. If the parabola is defined as $y = ax^2 + bx + c$ or as $y = a(x - h)^2 + k$, that distance is equal to $\dfrac{1}{4a}$ and is usually denoted by the variable p.

31

For the parabola $y = \dfrac{1}{4}x^2$, the value of p is $\dfrac{1}{4 \cdot \dfrac{1}{4}} = \dfrac{1}{1} = 1$. So the focus is

a point located 1 unit above the vertex, at (0, 1). The directrix is a horizontal line located 1 unit below the vertex, at $y = -1$.

| MATH FACTS |

If the vertex of a parabola is at (h, k), the focus is at $(h, k + p)$ and the directrix is at $y = k - p$, where $p = \dfrac{1}{4a}$.

Example 6

What are the coordinates of the focus and the equation of the directrix of the parabola defined by $y = \dfrac{1}{16}(x - 2)^2 - 5$?

Solution: Since this equation is in vertex form, the vertex is $(-(-2), -5) = (2, -5)$.

Since $a = \dfrac{1}{16}$, $p = \dfrac{1}{4 \cdot \dfrac{1}{16}} = 4$. The focus is at $(2, -5 + 4) = (2, -1)$, and the directrix is at $y = -5 - 4 = -9$.

Here is the graph of the parabola and its focus and directrix.

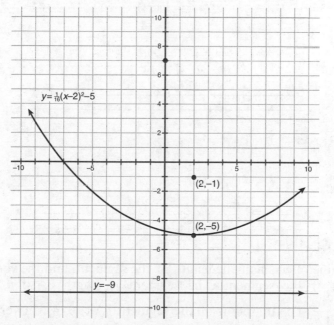

If the focus and directrix are known, it is possible to find the equation of the parabola related to that focus and directrix. The vertex is the midpoint of the vertical line segment connecting the focus to the directrix. The distance p is half the length of that vertical line segment, and the a-value for the quadratic equation equals $\dfrac{1}{4p}$.

If the focus is at $(-1, 4)$ and the directrix is at $y = -2$, a vertical line segment connecting $(-1, 4)$ to the line $y = -2$ would have length 6. The vertex would be the midpoint of that line segment, located at $(-1, 1)$. Half the length of the segment is the value of p, which is 3 for this example. The formula relating a and p is $a = \dfrac{1}{4p} = \dfrac{1}{4 \cdot 3} = \dfrac{1}{12}$.

Since you have the vertex and the a-value, the equation in vertex form is $y = \dfrac{1}{12}(x+1)^2 + 1$.

Example 7

What is the equation in standard form of a parabola that has its focus at $(0, 7)$ and a directrix of $y = 1$?

Solution: First sketch the focus and directrix on graph paper. Then draw the vertical segment joining the focus and the directrix.

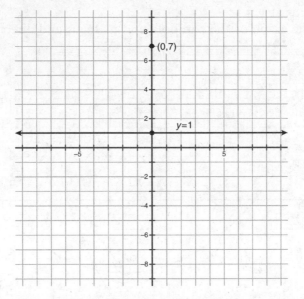

The value of p is half the length of the segment, and the vertex is the midpoint of the segment. The segment is length 6, so p is 3 and $a = \dfrac{1}{4p} = \dfrac{1}{12}$. The midpoint of the segment is at $(0, 4)$. In vertex form, the equation is $y = \dfrac{1}{12}(x-0)^2 + 4$. In standard form, it is $y = \dfrac{1}{12}x^2 + 4$.

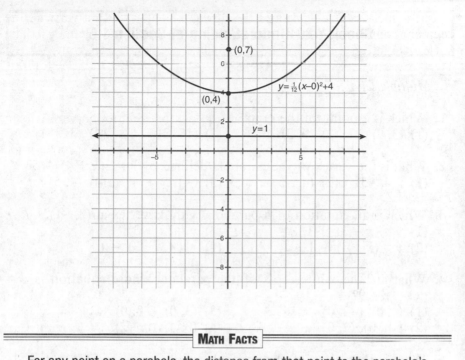

$$y = \tfrac{1}{12}(x-0)^2 + 4$$

(0,7)

(0,4)

y=1

MATH FACTS

For any point on a parabola, the distance from that point to the parabola's focus is the same as the distance from that point to the parabola's directrix. In the graph below, the congruent line segments are marked.

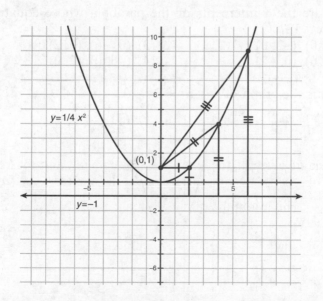

$y = 1/4\ x^2$

(0,1)

y=−1

Check Your Understanding of Section 1.5

A. *Multiple-Choice*

1. Which is a point on the graph of $y = 2x^2$?
   (1) $(3, 9)$ (2) $(9, 3)$ (3) $(18, 3)$ (4) $(3, 18)$

2. What is the vertex of the parabola defined by $y = 2(x + 4)^2 - 5$?
   (1) $(-4, -5)$ (2) $(4, -5)$ (3) $(-4, 5)$ (4) $(4, 5)$

3. Which is an equation of a parabola with its vertex at $(3, -1)$?
   (1) $y = (x - 3)^2 + 1$ (3) $y = (x + 3)^2 + 1$
   (2) $y = (x - 3)^2 - 1$ (4) $y = (x + 3)^2 - 1$

4. What are the x-intercepts of the parabola whose equation is $y = (x - 1)^2 - 9$?
   (1) $(4, 0), (-2, 0)$ (3) $(3, 0), (-3, 0)$
   (2) $(-4, 0), (2, 0)$ (4) $(0, 4), (0, -2)$

5. What is the vertex of the parabola whose equation is $y = x^2 + 4x - 12$?
   (1) $(-2, -16)$ (3) $(2, 0)$
   (2) $(-2, -24)$ (4) $(2, -16)$

6. What are the x-intercepts of the parabola whose equation is $y = x^2 - 4x - 12$?
   (1) $(6, 0), (-2, 0)$ (3) $(0, -6), (0, 2)$
   (2) $(0, 6), (0, -2)$ (4) $(-6, 0), (2, 0)$

7. Which is the graph of $y = x^2 + 2x - 3$?

(1)

36

(2)

(3)

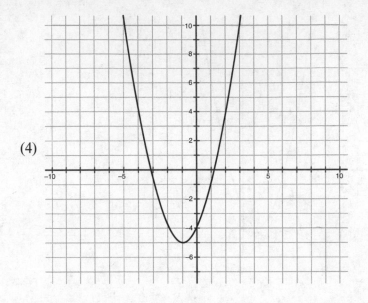

(4)

8. What is the equation of the directrix and the coordinates of the focus of the parabola $y = 3x^2$?

   (1) $y = -12$ and $(0, 12)$

   (2) $y = -\dfrac{1}{4}$ and $\left(0, \dfrac{1}{4}\right)$

   (3) $y = -\dfrac{1}{12}$ and $\left(0, \dfrac{1}{12}\right)$

   (4) $y = \dfrac{1}{12}$ and $\left(0, -\dfrac{1}{12}\right)$

9. What is the equation of the parabola that has the focus and directrix on this coordinate plane?

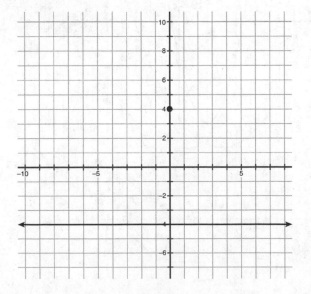

(1) $y = 12x^2$

(3) $y = \dfrac{1}{12}x^2$

(2) $y = x^2$

(4) $y = \dfrac{1}{12}x^2$

10. Which is the graph of $y = -\dfrac{1}{4}(x+3)^2 + 5$?

(1)

(2)

(3)

(4)

B. *Show how you arrived at your answers.*

1. Sketch the graph of $y = (x - 3)^2 - 1$, including the vertex, x-intercepts, and y-intercept.

2. $y = x^2 + bx + 7$ has a vertex at $(-4, -9)$. What is the value of b?

3. Convert $y = x^2 - 10x + 23$ into vertex form.

4. What is the equation of the parabola with its focus at (4, 3) and its directrix at $y = 1$?

5. If $m\overline{PD} = 8$, what is the length of $\overline{PF}$?

1.6 POLYNOMIAL GRAPHS

△
KEY IDEAS
◢◣

Graphs of cubic and other higher-degree polynomials can be sketched by knowing a few key points on the graph and having an understanding of the general shape of curves based on higher-degree polynomials. The *x*-intercepts and *y*-intercept are important as well as having an understanding of the *end behavior* at both ends of the curve.

Graphing Basic Cubic Equations

The graph of a quadratic function resembles a "U" (whether it is right-side up or upside-down). However, the graph of a cubic (3rd-degree) equation resembles a capital "N." The "N" can be either right-side up or upside-down.

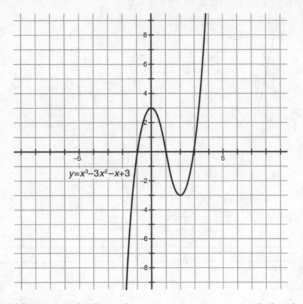

$y = x^3 - 3x^2 - x + 3$

It is useful if the cubic function you want to graph is already in factored form, such as $f(x) = (x + 1)(x - 1)(x - 3)$. If you were to multiply this out, the coefficient of the x^3-term would be +1, so the a-value is positive. The graph resembles a right-side-up "N."

The x-intercepts can be found by substituting 0 for y to get the equation $0 = (x + 1)(x - 1)(x - 3)$.

This equation can be solved by finding the x-values that would make any of the factors equal to zero.

$$x + 1 = 0 \quad \text{or} \quad x - 1 = 0 \quad \text{or} \quad x - 3 = 0$$

$$x = -1 \quad \text{or} \quad x = +1 \quad \text{or} \quad x = +3$$

So the x-intercepts are $(-1, 0)$, $(1, 0)$, and $(3, 0)$.

The y-intercept can be found by substituting $x = 0$ into the equation and solving for y.

$$y = (0 + 1)(0 - 1)(0 - 3) = +3$$

So the y-intercept is $(0, 3)$.

Plot the 4 points, and join them with a curve like a right-side-up "N."

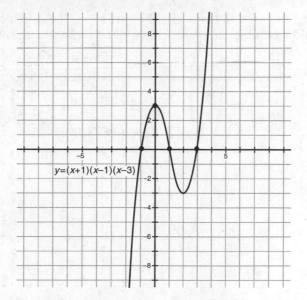

$y=(x+1)(x-1)(x-3)$

This equation can also be graphed with the graphing calculator.

Example 1

Sketch a graph of the function $f(x) = -\frac{1}{4}(x-2)(x-1)(x+4)$, including x-intercepts and y-intercept.

Solution: The x-intercepts can be found by using the equations $x - 2 = 0$, $x - 1 = 0$, and $x + 4 = 0$. The x-intercepts are $(2, 0)$, $(1, 0)$, and $(-4, 0)$. Since $y = -\frac{1}{4}(0-2)(0-1)(0+4) = -2$, the y-intercept is $(0, -2)$. Because there is a negative sign in front of the expression, the *a*-value is negative so this should resemble an upside-down "N." Draw the four points, and sketch the curve.

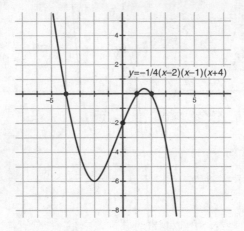

$y = -1/4(x-2)(x-1)(x+4)$

Graphing Polynomial Functions That Have Double Roots

A factored polynomial like $f(x) = (x - 2)(x - 2)(x + 3)$ has roots at 2, 2, and −3. Since there are two 2s, the graph of this function has only two *x*-intercepts: (2, 0) and (−3, 0). When there are two of the same root, it is called a *double root*. Graphing a polynomial that has double roots requires a special technique.

MATH FACTS

At a double root, the graph of the function "bounces off" the *x*-axis!

To graph $f(x) = (x - 2)(x - 2)(x + 3)$, first notice that the *a*-value is positive. So the graph will resemble a right-side-up "N."

The *x*-intercepts are at (2, 0) and (−3, 0). The *y*-coordinate of the *y*-intercept is at $y = (0 - 2)(0 - 2)(0 + 3) = 12$. So the *y*-intercept is (0, 12). Graph these three points.

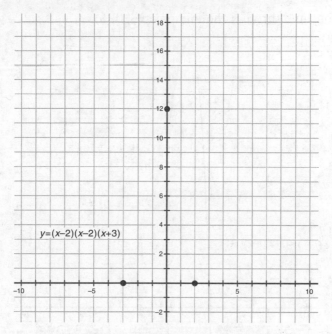

Since the graph should look like a right-side-up "N," start from the bottom left of the paper and draw toward the first x-intercept, $(-3, 0)$.

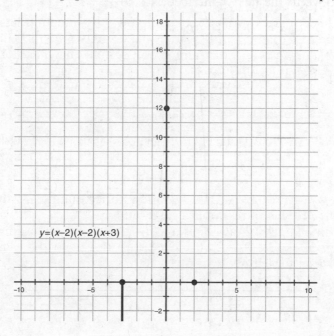

Then continue through the *y*-intercept, (0, 6).

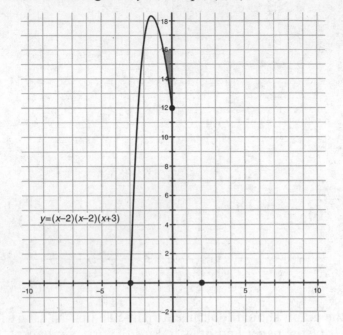

Then move to the next *x*-intercept, (2, 0).

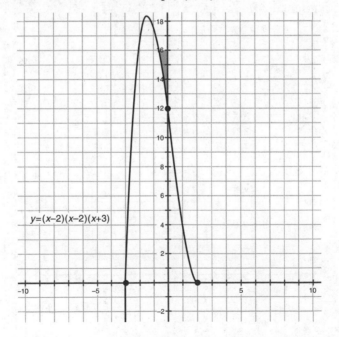

Since this is the double root, don't go through the x-axis. Instead, bounce off the axis and go back up to the right.

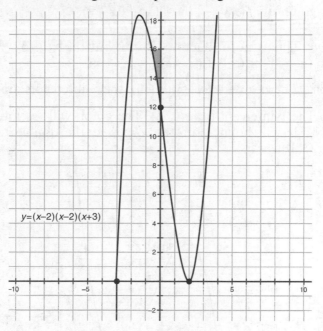

$y=(x-2)(x-2)(x+3)$

The graphing calculator agrees!

Example 2

Sketch the graph of $f(x) = \frac{1}{6}(x+4)(x^2 - 6x + 9)$.

Solution: This is not in factored form, but the $x^2 - 6x + 9$ can be factored to $(x - 3)^2$ or $(x - 3)(x - 3)$.

To graph $f(x) = \frac{1}{6}(x+4)(x-3)(x-3)$, the x-intercepts are at $(-4, 0)$ and $(3, 0)$. The y-intercept is at $(0, 6)$. The graph will resemble a right-side-up "N." It will bounce off the x-axis at the double root at the x-intercept $(3, 0)$.

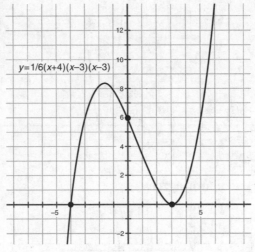

Example 3

If a, b, and c are all positive integers, which could be the graph of $f(x) = a(x - b)(x^2 - c^2)$?

(1)

48

(2)

(3)

(4)

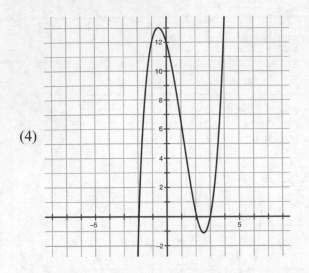

Solution: The $x^2 - c^2$ can be factored into $(x - c)(x + c)$, so the function becomes $f(x) = a(x - b)(x - c)(x + c)$. Since a is positive, the graph will resemble a right-side-up "N." The x-intercepts will be at $(b, 0)$, $(c, 0)$, and $(-c, 0)$.

An alternative is to make up three positive integers for a, b, and c, like $a = 1$, $b = 3$, $c = 2$. Then graph $y = 1(x - 3)(x^2 - 2^2)$ on the graphing calculator. See which answer choice most resembles it.

The solution is choice (4).

Check Your Understanding of Section 1.6

A. Multiple-Choice

1. Which could be the equation for this graph?

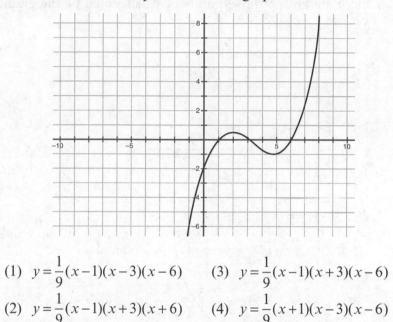

(1) $y = \dfrac{1}{9}(x-1)(x-3)(x-6)$ (3) $y = \dfrac{1}{9}(x-1)(x+3)(x-6)$

(2) $y = \dfrac{1}{9}(x-1)(x+3)(x+6)$ (4) $y = \dfrac{1}{9}(x+1)(x-3)(x-6)$

2. Which could be the equation for this graph?

(1) $y = \dfrac{1}{8}(x-2)^2(x-4)$ (3) $y = \dfrac{1}{16}(x-2)(x-4)^2$

(2) $y = -\dfrac{1}{4}(x-2)(x-4)$ (4) $y = -\dfrac{1}{8}(x-2)^2(x-4)$

3. What are the x-intercepts and y-intercept of $y = (x - 2)(x + 4)$ $(x + 5)$?
(1) $(-2, 0), (4, 0), (5, 0), (0, -40)$
(2) $(-2, 0), (4, 0), (5, 0), (0, 40)$
(3) $(2, 0), (-4, 0), (-5, 0), (0, -40)$
(4) $(2, 0), (-4, 0), (-5, 0), (0, 40)$

4. If a and b are both positive numbers, which could be the graph of $y = (x + a)^2(x - b)$?

(1)

(2)

(3)

(4)

5. Which equation could this be the graph of?

(1) $y = x - 2$ (3) $y = (x - 2)^3$

(2) $y = (x - 2)^2$ (4) $y = (x - 2)^4$

6. If this is the graph of $y = a(x - 2)(x - 1)(x + 3)$, what is the value of a?

(1) $\dfrac{1}{3}$ (2) $-\dfrac{1}{6}$ (3) $\dfrac{1}{6}$ (4) $-\dfrac{1}{3}$

7. If a and b are positive integers, which could be the graph of $y = (x - b)(x^2 + 2ax + a^2)$?

(1)

(2)

(3)

(4)

8. What are the *x*-intercepts of $y = x^3 - 2x^2 - 9x + 18$?
  (1) (2, 0), (3, 0), (4, 0)   (3) (−2, 0), (3, 0), (−3, 0)
  (2) (2, 0), (3, 0), (−3, 0)   (4) (−2, 0), (−9, 0), (18, 0)

9. What are the *x*-intercepts of the graph of $y = x^3 + 2x^2 - 8x$?
  (1) (2, 0), (−4, 0)   (3) (0, 0), (2, 0), (8, 0)
  (2) (0, 0), (−2, 0), (4, 0)   (4) (0, 0), (2, 0), (−4, 0)

10. What is the least number of *x*-intercepts an equation of the form $y = ax^3 + bx^2 + cx + d$ can have?
  (1) 0   (2) 1   (3) 2   (4) 3

B. *Show how you arrived at your answers.*

1. Sketch the graph of $y = \dfrac{1}{6}(x-2)(x+2)(x+3)$. Include all *x*- and *y*-intercepts.

2. What is the equation of this graph?

3. A graph of a cubic equation has a double root at *x* = 2 and a single root at *x* = −1. What could be the equation?

4. What is the main difference between the equations that represent these two graphs?

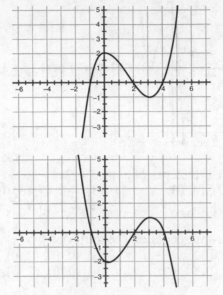

5. One of the x-intercepts of the cubic equation $y = x^3 + 4x^2 - 7x - 10$ is (2, 0). What are the other two x-intercepts?

Chapter Two

RATIONAL EXPRESSIONS AND EQUATIONS

2.1 ARITHMETIC WITH RATIONAL EXPRESSIONS

△
KEY IDEAS

A *rational expression* is a fraction that has a polynomial expression in the denominator. It often also has a polynomial expression in the numerator. An example of a rational expression is $\dfrac{x+3}{x^2+6x+8}$. Just like fractions involving integers, rational expressions can be simplified, reduced, multiplied, divided, added, and subtracted.

Reducing Rational Expressions

A *rational number* is a fraction, like $\dfrac{6}{10}$, that has an integer in both the numerator and the denominator. When both the numerator and denominator of a rational number are multiplied or divided by the same number, the result is a rational number that is equivalent to the original rational number.

$$\frac{6}{10}=\frac{2\cdot 6}{2\cdot 20}=\frac{12}{20}$$

$$\frac{6}{10}=\frac{6\div 2}{10\div 2}=\frac{3}{5}$$

$\dfrac{6}{10}, \dfrac{12}{20}$, and $\dfrac{3}{5}$ all represent the same number, 0.6.

Reducing a rational number is when the numerator and denominator are both divided by the same common factor. Factoring the numerator and denominator of a rational number makes it easier to reduce the fraction to *lowest terms*.

$$\frac{6}{10}=\frac{\cancel{2}\cdot 3}{\cancel{2}\cdot 5}=\frac{3}{5}$$

In the example above, we sometimes say that the 2s "canceled out."

Although dividing or multiplying the numerator and denominator by the same number results in an equivalent rational number, adding the same number to or subtracting the same number from the numerator and denominator does not produce an equivalent rational number. An example of this is $\dfrac{6}{10} \neq \dfrac{6-2}{10-2} = \dfrac{4}{8} = \dfrac{1}{2}$. For this reason, we sometimes say that we can cancel only numbers that are multiplied by both the numerator and the denominator, not numbers that are added to or subtracted from both the numerator and the denominator.

The value of a rational expression depends on the chosen value of x. The expression $\dfrac{5}{x+2}$ is equal to $\dfrac{5}{3}$ when $x = 1$. When $x = 2$, the value of the expression is $\dfrac{5}{4}$. When $x = -2$, the expression becomes $\dfrac{5}{0}$, which is undefined. Two rational expression are said to be equivalent if they evaluate to the same number for each x-value for which they are both defined.

One way to create a rational expression equivalent to another rational expression is to multiply the numerator and denominator by the same expression. When the numerator and denominator of the rational expression $\dfrac{5}{x+2}$ are multiplied by $x + 3$, the expression becomes $\dfrac{5(x+3)}{(x+2)(x+3)} = \dfrac{5x+15}{x^2+5x+6}$. This new expression is equivalent to the original one for every value of x except -3 since the new expression is undefined for $x = -3$. In contrast, the original expression is equal to -5 when $x = -3$.

Another way to create an equivalent rational expression is to reduce by finding a common factor in the numerator and denominator. Then cancel out those factors by dividing both the numerator and the denominator by that factor. This process is simplest when the numerator and denominator are already factored completely.

Example 1

Simplify the rational expression $\dfrac{3(x+5)}{(x+2)(x+5)}$.

Solution: The $(x + 5)$ factors can be canceled out, leaving the expression $\dfrac{3}{x+2}$.

If the expressions in the numerator and denominator of the rational expression are not already factored, factor them and cancel out equal factors.

Example 2

Simplify the rational expression $\dfrac{2x+8}{x^2+5x+4}$.

Solution: The numerator factors into $2(x+4)$. The denominator factors into $(x+4)(x+1)$. The rational expression then reduces to $\dfrac{2\cancel{(x+4)}}{\cancel{(x+4)}(x+1)} = \dfrac{2}{x+1}$.

MATH FACTS

The expression $b - a$ can be factored into $-1(a - b)$. This can be useful if in the numerator there is an expression like $x - 2$ and in the denominator there is an expression like $2 - x$. The expression in the denominator can be factored into $-1(x - 2)$. Then the $(x - 2)$ factors can be canceled out.

Example 3

Simplify the expression $\dfrac{x^2-9}{9-3x}$.

Solution: The numerator factors into $(x - 3)(x + 3)$, and the denominator factors into $3(3 - x)$. So the expression is $\dfrac{(x-3)(x+3)}{3(3-x)}$. Since there is an $x - 3$ in the numerator and a $3 - x$ in the denominator, rewrite the $3 - x$ as $-1(x - 3)$. The expression is now $\dfrac{\cancel{(x-3)}(x+3)}{-3\cancel{(x-3)}} = \dfrac{x+3}{-3}$.

Multiplying Rational Expressions

One way to multiply rational numbers is to multiply the numerators and the denominators of the factors and then reduce the solution.

$$\frac{22}{10} \cdot \frac{35}{21} = \frac{770}{210}$$

Reducing the fraction $\dfrac{770}{210}$ to lowest terms, however, can be complicated.

A shortcut is to factor each of the numbers in the two fractions, combine these factors into one numerator and one denominator, and then cancel any common factors.

$$\frac{22}{10} \cdot \frac{35}{21} = \frac{2 \cdot 11}{2 \cdot 5} \cdot \frac{7 \cdot 5}{7 \cdot 3} = \frac{\cancel{2} \cdot 11 \cdot \cancel{7} \cdot \cancel{5}}{\cancel{2} \cdot \cancel{5} \cdot \cancel{7} \cdot 3} = \frac{11}{3}$$

This same process can be applied to multiplying two rational expressions. By first factoring all the numerators and denominators, the factors to be canceled out are much more obvious.

$$\frac{11x + 33}{x^2 - x - 12} \cdot \frac{x^2 - 16}{3x + 12}$$

After all numerators and denominators have been factored using the different factoring patterns described in Chapter 1, the result is the following.

$$\frac{11 \cancel{(x+3)} \cancel{(x-4)} \cancel{(x+4)}}{\cancel{(x-4)} \cancel{(x+3)} 3 \cancel{(x+4)}} = \frac{11}{3}$$

Dividing Rational Expressions

Just like dividing rational numbers, rational expressions are divided by multiplying the first fraction by the reciprocal of the second fraction.

$$\frac{5}{(x+3)(x-2)} \div \frac{3(x+2)}{(x-2)(x+2)}$$

$$\frac{5}{(x+3)(x-2)} \cdot \frac{(x-2)(x+2)}{3(x+2)}$$

$$\frac{5 \cancel{(x-2)} \cancel{(x+2)}}{(x+3) \cancel{(x-2)} 3 \cancel{(x+2)}}$$

$$\frac{5}{3(x+3)}$$

Adding Rational Expressions

Like adding rational numbers, adding rational expressions requires finding a lowest common denominator. One of four situations typically occurs when adding rational numbers or expressions.

Situation 1: The expressions already have a common denominator.

Since $\dfrac{2}{7}$ and $\dfrac{3}{7}$ have a common denominator of 7, $\dfrac{2}{7} + \dfrac{3}{7} = \dfrac{2+3}{7} = \dfrac{5}{7}$.

Similarly, the rational expressions $\dfrac{x+2}{x+3}$ and $\dfrac{x+5}{x+3}$ have a common denominator of $x + 3$

$$\frac{x+2}{x+3} + \frac{x+5}{x+3} = \frac{(x+2)+(x+5)}{x+3} = \frac{x+2+x+5}{x+3} = \frac{2x+7}{x+3}$$

Situation 2: One denominator is a multiple of the other denominator.

For the fractions $\dfrac{2}{7}$ and $\dfrac{3}{14}$, the denominator 14 is a multiple of the denominator 7. The 7 needs to be multiplied by 2 to be equal to 14. Both the numerator and the denominator of the fraction $\dfrac{2}{7}$ must be multiplied by 2 to become the equivalent fraction $\dfrac{4}{14}$, which can then be added to $\dfrac{3}{14}$.

$$\frac{2}{7} + \frac{3}{14}$$
$$\frac{2 \cdot 2}{2 \cdot 7} + \frac{3}{14}$$
$$\frac{4}{14} + \frac{3}{14} = \frac{4+3}{14} = \frac{7}{14} = \frac{1}{2}$$

This situation can happen with rational expressions too.

$$\frac{2}{x+3} + \frac{3}{2x+6}$$

Factor the second denominator into $2(x + 3)$. This is equal to 2 times the other denominator.

$$\frac{2 \cdot 2}{2(x+3)} + \frac{3}{2(x+3)}$$
$$\frac{4}{2(x+3)} + \frac{3}{2(x+3)} = \frac{7}{2(x+3)} = \frac{7}{2x+6}$$

Situation 3: The denominators have no common factor.

If the denominators have no common factor, the lowest common denominator is the product of the two denominators.

This happens when adding two rational numbers like $\frac{2}{3} + \frac{3}{5}$. Since 3 and 5 have no common factor, the least common denominator is $3 \cdot 5 = 15$. Both denominators must become 15, and the numerators must change accordingly.

$$\frac{2}{3} + \frac{4}{5}$$

The numerator and the denominator of the first fraction get multiplied by the denominator of the second fraction. The numerator and the denominator of the second fraction get multiplied by the denominator of the first fraction.

$$\frac{5 \cdot 2}{5 \cdot 3} + \frac{3 \cdot 4}{3 \cdot 5} = \frac{10}{15} + \frac{12}{15} = \frac{22}{15}$$

With rational expressions, this situation frequently happens.

$$\frac{2}{x+1} + \frac{4}{x+3}$$

Since the expressions $(x + 1)$ and $(x + 3)$ have no common factor, the lowest common denominator is $(x + 1)(x + 3)$.

$$\frac{(x+3)2}{(x+3)(x+1)} + \frac{(x+1)4}{(x+1)(x+3)}$$

$$\frac{2x+6}{(x+1)(x+3)} + \frac{4x+4}{(x+1)(x+3)}$$

$$\frac{(2x+6)+(4x+4)}{(x+1)(x+3)} = \frac{2x+6+4x+4}{(x+1)(x+3)} = \frac{6x+10}{(x+1)(x+3)}$$

Situation 4: The two denominators share a common factor, but the larger denominator is not a multiple of the smaller one.

This situation is by far the most complicated one when adding rational numbers and rational expressions. This is seen with rational numbers when adding fractions like $\frac{7}{10}$ and $\frac{5}{6}$. The number 10 is not a multiple of 6, so this is not Situation 2. The two numbers have a common factor, 2, so this is not Situation 3. By factoring the two numbers $10 = 2 \cdot 5$ and $6 = 2 \cdot 3$, the least common multiple must have a factor of 2, 3, and 5, which makes it 30. Notice that if the two original denominators were multiplied, that would make a product of 60. This number could serve as a common denominator, but it is not the least common denominator.

$$\frac{7}{10} + \frac{5}{6}$$

$$\frac{3 \cdot 7}{3 \cdot 10} + \frac{5 \cdot 5}{5 \cdot 6}$$

$$\frac{21}{30} + \frac{25}{30} = \frac{46}{30}$$

This situation often happens with rational expressions.

$$\frac{7}{x^2 - 9} + \frac{5}{x^2 + 7x + 12}$$

Factor the denominators.

$$\frac{7}{(x-3)(x+3)} + \frac{5}{(x+3)(x+4)}$$

The common factor must have an $(x - 3)$, an $(x + 3)$, and an $(x + 4)$ in it. So the smallest factor is $(x - 3)(x + 3)(x + 4)$.

The first denominator needs an $(x + 4)$ to become $(x - 3)(x + 3)(x + 4)$. The second denominator needs an $(x - 3)$ to become $(x - 3)(x + 3)(x + 4)$.

$$\frac{(x+4)7}{(x+4)(x-3)(x+3)} + \frac{(x-3)5}{(x-3)(x+3)(x+4)}$$

$$\frac{7x+28}{(x+4)(x-3)(x+3)} + \frac{5x-15}{(x+4)(x-3)(x+3)}$$

$$\frac{(7x+28) + (5x-15)}{(x+4)(x-3)(x+3)}$$

$$\frac{7x+28 + 5x-15}{(x+4)(x-3)(x+3)}$$

$$\frac{12x + 13}{(x+4)(x-3)(x+3)}$$

Example 4

Simplify $\dfrac{3}{x^2 + x - 20} + \dfrac{2}{x^2 - 16}$.

Solution:

$$\frac{3}{(x+5)(x-4)}+\frac{2}{(x+4)(x-4)}$$

The common denominator is $(x + 5)(x - 4)(x + 4)$.

$$\frac{(x+4)3}{(x+5)(x-4)(x+4)}+\frac{(x+5)2}{(x+5)(x-4)(x+4)}$$

$$\frac{3x+12}{(x+5)(x-4)(x+4)}+\frac{2x+10}{(x+5)(x-4)(x+4)}$$

$$\frac{(3x+12)+(2x+10)}{(x+5)(x-4)(x+4)}$$

$$\frac{3x+12+2x+10}{(x+5)(x-4)(x+4)}$$

$$\frac{5x+22}{(x+5)(x-4)(x+4)}$$

Subtracting Rational Expressions

Subtracting rational expressions is nearly the same as adding them. An extra complication, which happens frequently on the Regents exam, is you must be careful distributing the negative sign through the parentheses of the second expression.

Even with two expressions that already have the same denominator, distributing the negative sign is an important issue.

Example 5

Simplify $\dfrac{3x+4}{x+2}-\dfrac{x-2}{x+2}$.

(1) $\dfrac{2x}{x+2}$

(2) $\dfrac{2x+2}{x+2}$

(3) $\dfrac{2x+4}{x+2}$

(4) $\dfrac{2x+6}{x+2}$

Solution: Choice (4) is correct.

$$\frac{(3x+4)-(x-2)}{x+2}$$

$$\frac{3x+4-x+2}{x+2}$$

$$\frac{2x+6}{x+2}$$

The most common incorrect answer is choice (2). This is what you would get if you distributed the negative sign through the second expression incorrectly as $3x + 4 - x - 2$. This error is one of the most common errors in all of high school math, so be careful while doing subtraction problems that involve polynomials!

Check Your Understanding of Section 2.1

A. Multiple-Choice

1. What is $\dfrac{x^2 - 4}{x^2 + x - 6}$ reduced to simplest terms?

   (1) $\dfrac{x+2}{x+3}$ (2) $\dfrac{-1}{x-2}$ (3) $\dfrac{x-2}{x-3}$ (4) $\dfrac{-4}{x-6}$

2. What is $\dfrac{2x+6}{x^2 - 2x - 15}$ reduced to simplest terms?

   (1) $\dfrac{6}{x^2 - 15}$ (2) $\dfrac{2x}{x^2 - 15}$ (3) $\dfrac{2}{x-5}$ (4) $\dfrac{6}{2x-3}$

3. What is $\dfrac{x^2 + 5x + 6}{x^2 - 9} \cdot \dfrac{5x - 10}{x^2 - 4}$ reduced to simplest terms?

   (1) $\dfrac{5}{x-3}$ (2) $\dfrac{5}{x+2}$ (3) $\dfrac{5}{x-2}$ (4) $\dfrac{5}{x+4}$

4. What is $\dfrac{12 - 3x}{5x - 5} \cdot \dfrac{x^2 - 1}{x^2 - 3x - 4}$ reduced to simplest terms?

   (1) $\dfrac{3}{5}$ (2) $\dfrac{12}{5}$ (3) $-\dfrac{3}{5}$ (4) $-\dfrac{12}{5}$

5. What is $\dfrac{4x+16}{x^2+5x+6} \div \dfrac{x^2-16}{5x+15}$ reduced to simplest terms?

(1) $\dfrac{20}{x+2}$

(3) $\dfrac{20}{(x+2)(x+3)}$

(2) $\dfrac{20}{x-4}$

(4) $\dfrac{20}{(x+2)(x-4)}$

6. When $\dfrac{x+2}{x-3} \cdot \dfrac{x+4}{x+5}$ is multiplied, which of the following does it have the same answer as?

(1) $\dfrac{x+2}{x-3} \div \dfrac{x+4}{x+5}$

(3) $\dfrac{x+2}{x-3} - \dfrac{x+4}{x+5}$

(2) $\dfrac{x+2}{x-3} + \dfrac{x+4}{x+5}$

(4) $\dfrac{x+2}{x-3} \div \dfrac{x+5}{x+4}$

7. What is $\dfrac{1}{x} + \dfrac{1}{x+2}$?

(1) $\dfrac{2}{2x+2}$

(2) $\dfrac{2x+2}{x(x+2)}$

(3) $\dfrac{1}{x(x+2)}$

(4) $\dfrac{2}{x(x+2)}$

8. What is $\dfrac{4}{(x+2)(x+3)} + \dfrac{3}{(x+2)(x-2)}$?

(1) $\dfrac{7x+1}{(x+2)(x-2)(x+3)}$

(3) $\dfrac{7x+3}{(x+2)(x-2)(x+3)}$

(2) $\dfrac{7}{(x+2)(x-2)(x+3)}$

(4) $\dfrac{7}{(x+2)^2(x-2)(x+3)}$

9. What is $\dfrac{5}{x-1} - \dfrac{3}{x+2}$?

(1) $\dfrac{2x+7}{(x-1)(x+2)}$

(3) $\dfrac{2}{-3}$

(2) $\dfrac{2x+13}{(x-1)(x+2)}$

(4) $\dfrac{2x+3}{(x-1)(x+2)}$

10. What is $\dfrac{x}{x^2+7x+12}-\dfrac{3}{x^2-9}$?

(1) $\dfrac{x^2-6x+12}{(x+3)(x-3)(x+4)}$

(3) $\dfrac{x^2-6x-12}{(x+3)(x-3)(x+4)}$

(2) $\dfrac{x-3}{7x+21}$

(4) $\dfrac{x-3}{7x+3}$

B. *Show how you arrived at your answers.*

1. Ethan says that $\dfrac{2x+5}{2}$ can be reduced to $x+5$. Braylon says this is not correct. Who is right and why?

2. James notices the following pattern:

$$\frac{1}{2}-\frac{1}{3}=\frac{1}{6}=\frac{1}{2\cdot 3}$$

$$\frac{1}{3}-\frac{1}{4}=\frac{1}{12}-\frac{1}{3\cdot 4}$$

$$\frac{1}{4}-\frac{1}{5}=\frac{1}{20}=\frac{1}{4\cdot 5}$$

He has a theory that, in general, $\dfrac{1}{x}-\dfrac{1}{x+1}=\dfrac{1}{x(x+1)}$. Prove that James is correct about his theory?

3. Simplify $\left(\dfrac{1}{x+h}-\dfrac{1}{x}\right)\div\left(\dfrac{x+h-x}{1}\right)$.

4. Talia simplified $\dfrac{5}{x+2}-\dfrac{2}{x-3}$ by this process:

$$\frac{5(x-3)-2(x+2)}{(x+2)(x-3)}=\frac{5x-15-2x+4}{(x+2)(x-3)}=\frac{3x-11}{(x+2)(x-3)}$$

There was an error in Talia's calculation. What was the error?

5. Fully simplify $\dfrac{x}{x+1}+\dfrac{x+1}{x+2}$.

6. The rational expression $\dfrac{x^3+12x^2+47x+60}{x^2-9}$ can be reduced. What is it in fully reduced form?

2.2 SOLVING RATIONAL EQUATIONS

A *rational equation* is an equation that contains at least one rational expression. Rational equations often require finding a common denominator for all the terms involved. The process for solving a rational equation often produces extra solutions that need to be rejected.

Simple Rational Equations

The simplest rational equations are ones that already have a common denominator.

$$\frac{2x+5}{(x+2)(x-3)} = \frac{13}{(x+2)(x-3)}$$

Since these have the same denominator already, the denominators can be ignored to create the equation with just the numerators.

$$2x + 5 = 13$$
$$\underline{-5 = -5}$$
$$2x = 8$$
$$\frac{2x}{2} = \frac{8}{2}$$
$$x = 4$$

With rational equations, it is necessary to check your answer in case the "ignoring" step somehow caused an incorrect answer to creep in. Simply plug your answer into the original problem.

$$\frac{2 \cdot 4 + 5}{(4+2)(4-3)} \stackrel{?}{=} \frac{13}{(4+2)(4-3)}$$
$$\frac{13}{6 \cdot 1} \stackrel{?}{=} \frac{13}{6 \cdot 1}$$
$$\frac{13}{6} \stackrel{\checkmark}{=} \frac{13}{6}$$

Solving Rational Equations with Cross Multiplication

Some rational equations have just one term on each side but the terms have different denominators.

$$\frac{2}{x+1} = \frac{4}{x+4}$$

The quickest way to solve an equation of this form is by *cross multiplication*. To cross multiply, make a new equation that has the product of one numerator with the other denominator on each side of the equal sign.

$$2(x+4) = 4(x+1)$$

$$
\begin{array}{rcl}
2x+8 &=& 4x+4 \\
-2x &=& -2x \\
\hline
8 &=& 2x+4 \\
-4 &=& -4 \\
\hline
4 &=& 2x \\
\frac{4}{2} &=& \frac{2x}{2} \\
2 &=& x
\end{array}
$$

To check if this is correct, substitute $x = 2$ into the original equation to see if the left side and the right side evaluate to the same number.

$$\frac{2}{2+1} \overset{?}{=} \frac{4}{2+4}$$

$$\frac{2}{3} \overset{\checkmark}{=} \frac{4}{6}$$

MATH FACTS

Cross multiplication is a shortcut for making both sides of the equal sign have a common denominator. You then ignore the denominators and make an equation out of the numerators. If $\frac{a}{b} = \frac{c}{d}$, use cross multiplication to get the common denominator bd: $\frac{ad}{bd} = \frac{bc}{bd}$ or $ad = bc$.

Solving Rational Equations with Multiple Terms

If the rational equation has more than one term on either side of the equal sign (normally the left side), those terms need to be combined. Then the resulting equation can be solved using the methods mentioned earlier in this section.

$$\frac{x-1}{x+2} + \frac{x}{x+1} = \frac{6x+5}{x^2+3x+2}$$

$$\frac{x-1}{x+2} + \frac{x}{x+1} = \frac{6x+5}{(x+2)(x+1)}$$

$$\frac{(x+1)(x-1)}{(x+1)(x+2)} + \frac{(x+2)x}{(x+2)(x+1)} = \frac{6x+5}{(x+2)(x+1)}$$

$$\frac{x^2-1}{(x+1)(x+2)} + \frac{x^2+2x}{(x+2)(x+1)} = \frac{6x+5}{(x+2)(x+1)}$$

$$\frac{2x^2+2x-1}{(x+2)(x+1)} = \frac{6x+5}{(x+2)(x+1)}$$

$$2x^2+2x-1 = 6x+5$$
$$\underline{-6x-5 = -6x-5}$$
$$2x^2-4x-6 = 0$$
$$2(x^2-2x-3) = 0$$
$$2(x-3)(x+1) = 0$$

$$x-3=0 \quad \text{or} \quad x+1=0$$
$$x=3 \quad \text{or} \quad x=-1$$

The process of ignoring the denominator can result in fake answers. If you substitute $x = 3$ into the original equation, it becomes $\frac{2}{5} + \frac{3}{4} = \frac{23}{20}$, which is true. However, if you substitute $x = -1$ into the original equation, the denominator of the second term becomes 0, so the expression is undefined. The $x = -1$ solution must be rejected, leaving only the solution $x = 3$.

A question like this where a solution must be rejected is common on the Algebra II Regents.

Example 1

Solve for all values of x that satisfy the equation $\frac{x}{x+1} + \frac{x-1}{x+4} = \frac{8x+5}{(x+1)(x+4)}$.

Solution: The common denominator for the two rational expressions on the left side of the equal sign is $(x + 1)(x + 4)$.

$$\frac{x}{x+1} + \frac{x-1}{x+4} = \frac{8x+5}{x^2+5x+4}$$

$$\frac{(x+4)x}{(x+4)(x+1)} + \frac{(x+1)(x-1)}{(x+1)(x+4)} = \frac{8x+5}{(x+1)(x+4)}$$

$$\frac{x^2+4x}{(x+4)(x+1)} + \frac{x^2-1}{(x+1)(x+4)} = \frac{8x+5}{(x+1)(x+4)}$$

$$\frac{(x^2+4x)+(x^2-1)}{(x+4)(x+1)} = \frac{8x+5}{(x+1)(x+4)}$$

$$\frac{x^2+4x+x^2-1}{(x+4)(x+1)} = \frac{8x+5}{(x+1)(x+4)}$$

$$\frac{2x^2+4x-1}{(x+4)(x+1)} = \frac{8x+5}{(x+1)(x+4)}$$

Since the two rational expressions have the same denominator, that denominator can be ignored.

$$\begin{aligned} 2x^2 + 4x - 1 &= 8x + 5 \\ -8x - 5 &= -8x - 5 \\ \hline 2x^2 - 4x - 6 &= 0 \end{aligned}$$

$$2(x^2 - 2x - 3) = 0$$

$$2(x-3)(x+1) = 0$$

$$x - 3 = 0 \quad \text{or} \quad x + 1 = 0$$

$$x = 3 \quad \text{or} \quad x = -1$$

Although $x = 3$ does make the original equation true, $\frac{4}{5} + \frac{3}{8} = \frac{37}{40}$, the $x = -1$ makes two of the rational expressions undefined. So $x = -1$ has to be rejected. The only solution is $x = 3$.

Rate Word Problems

Some real-world word problems involving rates can be solved using rational equations.

Example 2

Arianna can complete a coding project in 3 hours. Elijah can complete the same coding project in 2 hours. How long will it take them if they work together?

Solution: Arianna's rate is $\dfrac{1}{3}$ project per hour. Elijah's rate is $\dfrac{1}{2}$ project per hour. Their rate together is $\dfrac{1}{x}$ project per hour. Write an equation to solve the problem.

$$\frac{1}{2} + \frac{1}{3} = \frac{1}{x}$$
$$\frac{3x}{6x} + \frac{2x}{6x} = \frac{6}{6x}$$
$$3x + 2x = 6$$
$$5x = 6$$
$$x = \frac{6}{5} = 1\frac{1}{5} = 1.2$$

Working together would take them 1.2 hours.

Example 3

Two cars are driving from Washington, D.C. to New York, a distance of 300 miles. The first car went 10 miles per hour faster than the second car and got to New York 1 hour sooner. How long did it take each car to make the trip?

Solution: If x is the first car's time, then $\dfrac{300}{x}$ is the car's speed. Since the second car took one more hour, $x + 1$ is the second car's time and $\dfrac{300}{x+1}$ is the second car's speed. Since the first car's speed is 10 mph more than the second car's speed, the equation is $\dfrac{300}{x+1} + 10 = \dfrac{300}{x}$.

The solution is $x = 5$. So the first car took 5 hours, and the second car took $x + 1 = 6$ hours.

Check Your Understanding of Section 2.2

A. *Multiple-Choice*

1. Solve for x.

$$\frac{3x+5}{x-2} = \frac{x+13}{x-2}$$

(1) 2   (2) 3   (3) 4   (4) 5

2. Solve for x.

$$\frac{5x-3}{2x^2-3x} = \frac{2x+12}{2x^2-3x}$$

(1) 2   (2) 3   (3) 4   (4) 5

3. Solve for x.

$$\frac{x+1}{x+2} = \frac{x+4}{x+6}$$

(1) 0   (2) 1   (3) 2   (4) 3

4. Solve for x.

$$\frac{x+1}{x+2} = \frac{2x+2}{x+7}$$

(1) $x = 1$ (3) $x = -1$ or $x = 3$
(2) $x = 3$ (4) No solutions

5. Solve for x.

$$\frac{3}{x+2} + \frac{2}{x-3} = \frac{15}{(x+2)(x-3)}$$

(1) 2   (2) 3   (3) 4   (4) 5

6. Solve for x.

$$\frac{1}{x-4} + \frac{3}{x-1} = \frac{7}{x^2-5x+4}$$

(1) 4   (2) 5   (3) 6   (4) 7

7. Solve for x.

$$\frac{x}{x-3} + \frac{2}{x+1} = \frac{x+2}{x^2-2x-3}$$

(1) $x = 2$ or $x = -4$ (3) $x = 2$

(2) $x = -2$ or $x = 4$ (4) $x = -4$

8. Solve for x.

$$\frac{3}{x-4} - \frac{2}{x+5} = \frac{13}{(x-4)(x+5)}$$

(1) -10 (2) -6 (3) 10 (4) 6

9. Solve for x.

$$\frac{6}{x+4} - \frac{4}{x+2} = \frac{12}{x^2+6x+8}$$

(1) -6 (2) 6 (3) -8 (4) 8

10. Solve for x.

$$\frac{x+2}{x-4} - \frac{3}{x+1} = \frac{23}{x^2-3x-4}$$

(1) $x = 3$ or $x = -3$ (3) $x = -3$

(2) $x = 3$ (4) $x = 4$ or $x = -2$

B. *Show how you arrived at your answers.*

1. A group of people contribute equal amounts of money to get a $24 gift for a friend. If two more people contributed and they all paid equal amounts, they would each pay $1 less. How many people were there originally?

2. Leah solves $\dfrac{x}{(x-2)(x+2)} - \dfrac{3}{(x+2)(x-4)} = \dfrac{-4}{(x-2)(x+2)(x-4)}$ and gets solutions of $x = 5$ and $x = 2$. Paxton says that just $x = 5$ is the solution. Who is correct?

3. Kevin and Noelle drive 240 miles to Boston to watch a Yankees vs. Red Sox game. On the way home, they drive 10 mph slower and get home 48 minutes later than they would have had they driven the same speed as before. How fast did they drive from New York to Boston?

4. What is the solution to $\dfrac{1}{x+2} + \dfrac{2}{x+3} + \dfrac{5}{x+4} = \dfrac{178}{(x+2)(x+3)(x+4)}$?

5. Solve for x.

$$\frac{x+3}{x-1} - \frac{x-1}{x+2} = \frac{26}{(x-1)(x+2)}$$

2.3 GRAPHING RATIONAL FUNCTIONS

KEY IDEAS

Just like polynomial, exponential, and radical functions, rational functions can be graphed. The graphs of rational functions usually include horizontal and/or vertical *asymptotes*. Asymptotes are like invisible lines that curves get closer and closer to. The graph of a rational function can help solve certain rational equations.

Vertical and Horizontal Asymptotes

Below is the graph of the rational function $R(x) = \dfrac{1}{x-2}$.

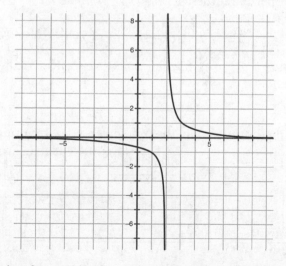

A characteristic of a graph of a rational function is that it often has one or more *vertical asymptotes*. A vertical asymptote is a vertical line that the graph never touches or crosses. In this graph, the vertical asymptote is the vertical line at $x = 2$. If you evaluate $R(2) = \dfrac{1}{2-2} = \dfrac{1}{0}$, it is undefined. For numbers very close to 2, such as 2.000001, the function produces a very high number. For x = 2.000001, $R(2.000001) = \dfrac{1}{2.000001 - 2} = \dfrac{1}{0.000001} = 1,000,000$.

Some graphs of rational functions also have a *horizontal asymptote*. Like the vertical asymptote, it is a horizontal line that, usually for very high and very low x-values, the curve gets closer and closer to without touching or crossing. In the graph for $R(x) = \dfrac{1}{x-2}$, there is a horizontal asymptote at $y = 0$, the x-axis.

The asymptotes have an effect on the domain and range of the function. For $R(x) = \dfrac{1}{x-2}$, the domain is everything except $x = 2$ and the range is everything except $y = 0$.

When graphing rational functions on the graphing calculator, be sure to use parentheses carefully or to use the fraction template.

Graph the more complicated function $R(x) = \dfrac{x-5}{(x+1)(x+4)}$ using a graphing calculator.

For the TI-84:

If you are in classic mode, enter the function into the Y= screen. Make sure to use proper parentheses.

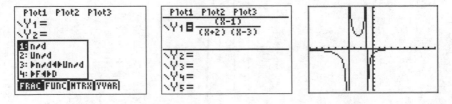

If you are in mathprint mode, the Y= screen is more intuitive.

For the TI-Nspire:

After the f1(x) in the entry line of the graphing scratchpad, press the [template] button and select the fraction template. Enter x 5 into the numerator and $(x + 1)(x + 4)$ into the denominator.

Checking to See if Two Rational Expressions Are Equivalent Using the Graphing Calculator

For a multiple-choice question that asks you to simplify a rational expression, the graph of the original expression can be compared to the graphs of the answer choices.

Example

Simplify $\dfrac{3x+4}{x+2} - \dfrac{x-2}{x+2}$.

(1) $\dfrac{2x}{x+2}$ (3) $\dfrac{2x+4}{(x+2)}$

(2) $\dfrac{2x+2}{x+2}$ (4) $\dfrac{2x+6}{x+2}$

Solution: This example was done earlier in the chapter with algebra but can also be done with the graphing calculator. By creating the graph of the original expression and the graph of the four answer choices, only choice (4) has the same graph as the original. Just look at the TI-84 screen shots below.

Graphs of choices:

Solving Rational Equations with the Graphing Calculator

The equation $\dfrac{2}{x+1} = \dfrac{4}{x+4}$, which was solved with algebra earlier in this chapter, can also be solved with the intersect feature of the graphing calculator. To solve a rational equation with the graphing calculator, graph both the expression on the left side of the equal sign and the expression on the right side of the equal sign. The x-coordinate of the intersection point is the solution to the equation.

Since the two expressions intersect at a point with an x-coordinate of 2, the solution is $x = 2$.

Check Your Understanding of Section 2.3

A. Multiple-Choice

1. What are the asymptotes of the graph of $R(x) - \dfrac{3}{x+4}$?

   (1) $x = 4,\ y = 0$ (3) $x = 0,\ y = 4$
   (2) $x = -4,\ y = 0$ (4) $x = 0,\ y = -4$

2. Which of the following could be the equation for this graph?

   (1) $R(x) = \dfrac{12}{x+3}$ (3) $R(x) = \dfrac{x-3}{12}$

   (2) $R(x) = \dfrac{x+3}{12}$ (4) $R(x) = \dfrac{12}{x-3}$

3. Which of the following is the graph of $y = \dfrac{8}{x+2}$?

(1)

(2)

(3)

(4)

4. Use a graphing calculator to determine which equation has the same graph as $y = \dfrac{2}{x+1} + \dfrac{3}{x}$.

(1) $y = \dfrac{5x+3}{x(x+1)}$ (3) $y = \dfrac{5x+3}{2x+1}$

(2) $y = \dfrac{5}{2x+1}$ (4) $y = \dfrac{5}{x(x+1)}$

5. Which of the following has the same graph as $y = \dfrac{x^2 + 5x + 6}{x^2 - 4}$?

(1) $y = \dfrac{x+2}{x-2}$ (3) $y = 1 - \dfrac{5}{4}x + 6$

(2) $y = \dfrac{x+3}{x-2}$ (4) $y = \dfrac{x-3}{x+2}$

EXPONENTIAL AND LOGARITHMIC EXPRESSIONS AND EQUATIONS

3.1 PROPERTIES OF EXPONENTS

An expression like 2^5 is called an *exponential expression*. It has a *base*, which in this case is the number 2, and an *exponent*, which in this case is the number 5. For positive integer exponents, the expression can be evaluated by multiplying the base by itself the number of times of the exponent.

The expression 2^5 is equal to 32 because $2 \cdot 2 \cdot 2 \cdot 2 \cdot 2 = 32$. This is very different from $2 \cdot 5 = 10$. So $2^5 \neq 2 \cdot 5$.

Multiplying Exponential Expressions

When two exponential expressions that have the same base are multiplied, the product can be written as an exponential expression that also has that base but whose exponent is the sum (not the product) of the exponents. If the bases are different, there is no simple way to multiply the expressions.

Example 1

Simplify $2^3 \cdot 2^5$.

    (1) 2^{15} (2) 4^{15} (3) 2^8 (4) 4^8

    Solution: Choice (3) is correct: 2^8. To see why the rule is to keep the base and add, rather than multiply, the two exponents, simplify $2^3 \cdot 2^5$ the long way.

$$2^3 \cdot 2^5 = (2 \cdot 2 \cdot 2) \cdot (2 \cdot 2 \cdot 2 \cdot 2 \cdot 2) = 2 \cdot 2 \cdot 2 \cdot 2 \cdot 2 \cdot 2 \cdot 2 \cdot 2 = 2^8$$

Dividing Exponential Expressions

When two exponential expressions that have the same base are divided, the product can be written as an exponential expression that also has that base but whose exponent is the difference of the exponents. If the bases are different, there is no simple way to divide the expressions.

Example 2

Simplify $2^{12} \div 2^4$.

    (1) 1^3 (2) 1^8 (3) 2^3 (4) 2^8

Solution: Choice (4) is correct: 2^8. To see why the rule is to keep the base and subtract the exponent, write the expression as a fraction and reduce.

$$\frac{(2^{12})}{2^4} = \frac{2 \cdot 2 \cdot 2 \cdot 2 \cdot 2 \cdot 2 \cdot 2 \cdot 2 \cdot \cancel{2} \cdot \cancel{2} \cdot \cancel{2} \cdot \cancel{2}}{\cancel{2} \cdot \cancel{2} \cdot \cancel{2} \cdot \cancel{2}} = 2 \cdot 2 \cdot 2 \cdot 2 \cdot 2 \cdot 2 \cdot 2 \cdot 2 = 2^8$$

Raising a Power to a Power

To raise an exponential expression to a power, keep the same base as in the original expression. Multiply the two exponents to find the new exponent.

Example 3

Simplify $(2^3)^2$.

Solution: Keep the base and multiply the two exponents. The answer is 2^6. To see why, expand.

$$(2^3)^2 = (2^3)(2^3) = 2^{3+3} = 2^{3 \cdot 2} = 2^6$$

Example 4

Which expression is equivalent to 1.07^x?

    (1) 1.07^{12x} (2) $(1.07^{12})^x$ (3) 1.12^{7x} (4) $\left(1.07^{\frac{1}{12}}\right)^{12x}$

Solution: Choice (4) is correct. To raise an expression that has an exponent in it to a power, multiply the two exponents. In this case $\frac{1}{12} \cdot 12x = 1x = x$. So choice (4) is equivalent to 1.07^x.

The power rule for exponents can also be used in reverse. If there is an expression like 2^{3x}, it can be converted to $(2^3)^x = 8^x$. The expressions 2^{3x} and 8^x are equivalent.

When the exponent is not a positive integer, things get a bit more complicated. For instance 2^0 is *not* equal to 0, as you might expect. Instead, 2^0 is equal to 1. Negative integer exponents may also seem unusual, even impossible. However, 2^{-3} is not undefined or even -8. Test it on a calculator. The answer is 0.125!

These unusual situations arise from the rule for dividing exponential expressions. $2^5 \div 2^5 = 2^{5-5} = 2^0$. However, any number divided by itself is equal to 1, so $2^0 = 1$.

$2^1 \div 2^4 = 2^{1-4} = 2^{-3}$. However, $\dfrac{2^1}{2^4}$ reduces to $\dfrac{1}{2^3} = \dfrac{1}{8} = 0.125$.

MATH FACTS

There are three main properties of exponents.

$$x^a \cdot x^b = x^{a+b}$$

$$\frac{x^a}{x^b} = x^{a-b}$$

$$(x^a)^b = x^{ab}$$

MATH FACTS

Any number (besides 0) raised to the 0 power is equal to 1. Any number (besides 0) raised to a negative power is equal to the reciprocal of that number raised to the positive version of that power.

$$b^0 = 1$$

$$b^{-n} = \frac{1}{b^n}$$

Example 5

What is the value of 3^{-4} rounded to the nearest thousandth?

Solution:

$$3^{-4} = \frac{1}{3^4} = \frac{1}{81} \approx 0.012$$

Example 6

Write an expression equivalent to $\dfrac{x^{-2}}{y^{-3}}$ that uses just positive exponents.

Solution: Remember that $x^{-2} = \dfrac{1}{x^2}$ and $y^{-3} = \dfrac{1}{y^3}$.

$$\frac{x^{-2}}{y^{-3}} = \frac{\dfrac{1}{x^2}}{\dfrac{1}{y^3}} = \frac{1}{x^2} \cdot \frac{y^3}{1} = \frac{y^3}{x^2}$$

Notice how the x^{-2} moved to the denominator as x^2 while the y^{-3} moved to the numerator as a y^3. In general, when there is a negative exponent, the expression containing that exponent can move from numerator to denominator or from denominator to numerator with the negative exponent changing into a positive exponent.

To raise a number to a fractional power, like $4^{\frac{3}{2}}$, first notice that

$4^1 = 4$ and $4^2 = 16$. Since the exponent $\dfrac{3}{2}$ is between 1 and 2, the answer should be something between 4 and 16. However, it is not halfway between 4 and 16, which is 10 as you might expect. Instead, if you test it on a calculator, you get 8!

==============================
MATH FACTS
==============================

To raise a base to a fractional power $\dfrac{n}{d}$, take the *d*th root of the number and raise it to the *n*th power.

$$b^{\frac{n}{d}} = \left(\sqrt[d]{b}\right)^n$$

_____

Example 7

What is the value of $8^{\frac{1}{3}}$?

Solution:

$$8^{\frac{1}{3}} = \left(\sqrt[3]{8}\right)^1 = 2^1 = 2$$

Notice that when the numerator of the exponent is 1, the solution just becomes the *d*th root of the base where *d* is the denominator. This is because the multiplication rule for exponents says that

$8^{\frac{1}{3}} \cdot 8^{\frac{1}{3}} \cdot 8^{\frac{1}{3}} = 8^{\frac{1}{3} + \frac{1}{3} + \frac{1}{3}} = 8^1 = 8$, while the definition of cube root says that

$\sqrt[3]{8} \cdot \sqrt[3]{8} \cdot \sqrt[3]{8} = 8$. So $8^{\frac{1}{3}} = \sqrt[3]{8}$.

Example 8

What is the value of $8^{\frac{5}{3}}$?

Solution:

$$8^{\frac{5}{3}} = \left(\sqrt[3]{8}\right)^5 = 2^5 = 32$$

The Distributive Property for Exponents

If an expression in parentheses is the product of numbers and/or variables, the entire expression in the parentheses can be raised to a power by raising each of the factors to that power and multiplying them together. For example, $(2 \cdot 5)^3$ can be calculated by solving $2^3 \cdot 5^3 = 8 \cdot 125 = 1,000$. This is also the solution if you simplified inside the parentheses first: $10^3 = 1,000$. This property is needed when variables are involved.

Example 9

Simplify $(3x^2y^3)^2$.

Solution:

$$3^2(x^2)^2(y^3)^2 = 9x^{2 \cdot 2}y^{3 \cdot 2} = 9x^4y^6$$

Example 10

Simplify the expression $\dfrac{\sqrt[3]{x^5}}{(x^2)^{\frac{1}{3}}}$

Solution: First express the numerator and the denominator as x raised to a power represented by a single number.

$$\sqrt[3]{x^5} = x^{\frac{5}{3}}$$

$$(x^2)^{\frac{1}{3}} = x^{2 \cdot \frac{1}{3}} = x^{\frac{2}{3}}$$

$$\frac{\sqrt[3]{x^5}}{(x^2)^{\frac{1}{3}}} = \frac{x^{\frac{5}{3}}}{x^{\frac{2}{3}}} = x^{\frac{5}{3} - \frac{2}{3}} = x^1$$

Check Your Understanding of Section 3.1

A. *Multiple-Choice*

1. Simplify $3^4 \cdot 3^5$.
   (1) 3^{20} (2) 3^9 (3) 9^9 (4) 9^{20}

2. Simplify $(5^2)^3$.
   (1) 5^8 (2) 5^6 (3) 5^5 (4) 5^9

3. What is a^{-4}?
   (1) $-(a^4)$ (2) $a^{\frac{1}{4}}$ (3) $\dfrac{1}{a^4}$ (4) $\sqrt[4]{a}$

4. What is $a^{\frac{3}{4}}$?
   (1) $\sqrt[3]{a^4}$ (2) $a^{-\frac{3}{4}}$ (3) $a^{-\frac{4}{3}}$ (4) $\sqrt[4]{a^3}$

5. Simplify $(2x^3y^4)^3$.
   (1) $8x^{27}y^{64}$ (2) $6x^9y^{12}$ (3) $8x^9y^{12}$ (4) $6x^{27}y^{64}$

6. What is $8^{-\frac{2}{3}}$?
   (1) $\dfrac{1}{4}$ (2) $-\dfrac{1}{4}$ (3) 4 (4) -4

7. What is 5^0?
   (1) 0 (2) 1 (3) 5 (4) Undefined

8. What is $\dfrac{8x^4y^2}{2x^2y^5}$?
   (1) $\dfrac{6x^2}{y^3}$ (2) $\dfrac{6x^2}{y^2}$ (3) $\dfrac{4x^2}{y^3}$ (4) $\dfrac{4x^3}{y^2}$

9. What is $7^{\frac{1}{2}} \cdot 7^{\frac{3}{2}}$?
   (1) 47 (2) 48 (3) 49 (4) 50

10. What is $11^{-4} \cdot 11^4$?
   (1) 0 (2) 1 (3) 11 (4) 121

B. *Show how you arrived at your answers.*

1. Ashlynn says that $5^3 \cdot 5^4 = 5^7$. Colin says that it is equal to 5^{12}. Who is right and why?

2. In 5th grade, Charles learned that $5.2 \times 10^{-4} = 0.00052$. Show how the properties of negative exponents justify this answer.

3. If $2^{10} = 1,024$, how can you quickly calculate 2^{11} if your calculator does not have an exponent key?

4. If $9^1 = 9$ and $9^2 = 81$, what is the value of $9^{1.5}$? (Hint: change 1.5 into an improper fraction.)

5. What is the value of $8^{-2} \cdot 25^{\frac{3}{2}} \cdot 27^{-\frac{2}{3}}$?

3.2 SOLVING EXPONENTIAL EQUATIONS BY GUESS AND CHECK OR BY GRAPHING

KEY IDEAS

An *exponential equation* is one in which the variable is an exponent. An example is the equation $2^x = 32$. Some exponential equations have integer solutions, some have rational solutions (fractions), and some have irrational solutions. One way to solve exponential equations is through guess and check. Another way is to use the intersect feature of a graphing calculator.

Solving Exponential Equations with Guess and Check

To solve $2^x = 32$ with guess and check, first think of a number that might be reasonable. Suppose you guess that $x = 10$. To check, calculate 2^{10}, which is not 32 but, 1,024. So 10 was too big of a guess. Since the answer to this question is an integer, after enough guessing and checking, you will come to the answer $x = 5$ since $2^5 = 32$.

Example 1

Solve $3^x = 729$ by guess and check.

Solution: Since $3^6 = 729$, the solution is $x = 6$.

When a positive integer is raised to a negative power, the result is a number between 0 and 1. In an exponential equation where the number after the equals sign is between 0 and 1, the exponent is negative.

Example 2

Which value of x makes $5^x = \dfrac{1}{125}$?

(1) -2 (2) -3 (3) $\dfrac{1}{2}$ (4) $\dfrac{1}{3}$

Solution: Since $\dfrac{1}{125}$ is between 0 and 1, the answer must be negative.

Since $5^{-2} = \dfrac{1}{5^2} = \dfrac{1}{25}$, the denominator is too small. Since $5^{-3} = \dfrac{1}{5^3} = \dfrac{1}{125}$, the answer is $x = -3$, which is choice (2).

When the solution to an exponential equation is not an integer, guess and check can be used to find an approximate answer.

Example 3

To the nearest tenth, what is the solution to the equation $2^x = 42$?

Solution: Since $2^5 = 32$, which is too small, and since $2^6 = 64$, which is too big, the answer is between 5 and 6. Test the numbers 5.1, 5.2, 5.3, and so on until you find a number that is close to 42.

$$2^{5.1} \approx 34$$
$$2^{5.2} \approx 37$$
$$2^{5.3} \approx 39$$
$$2^{5.4} \approx 42$$

Sometimes it is possible to get an exact solution even when the solution is not an integer. In those cases, the exponent is a rational number (a fraction).

Example 4

What is the exact solution to $4^x = 32$?

Solution: Since $4^2 = 16$ and $4^3 = 64$, the answer is between 2 and 3. Since $32 = 2^5$ and $2 = \sqrt{4} = 4^{\frac{1}{2}}, 32 = \left(4^{\frac{1}{2}}\right)^5 = 4^{\frac{1}{2} \cdot 5} = 4^{\frac{5}{2}}$. So the solution is $x = \dfrac{5}{2}$.

Exponential equations with exponents on both sides of the equal sign are easiest to solve when the two bases are the same.

Example 5

What value of x makes $3^x = 3^7$?

Solution: Since the bases are the same, the exponents must be equal. So $x = 7$.

Example 6

What value of x makes $3^{2x+3} = 3^7$?

Solution: Since the bases are the same, the exponents must be equal. This means that $2x + 3 = 7$. This can be solved with algebra: $x = 2$.

If the bases are different, they can sometimes be converted into the same base and then solved by equating the exponents.

Example 7

What value of x makes $3^{2x-4} = 9^4$?

Solution: Since 9 can be written as 3^2, the right side of the equation becomes $(3^2)^4$. This is equivalent to 3^8 because powers can be raised to powers by multiplying the exponents.

The equation is now $3^{2x-4} = 3^8$.

$$2x - 4 = 8$$
$$x = 6$$

Graphs of Two-Variable Exponential Equations

An equation of the form $y = a \cdot b^x + c$ where a, b, and c are numbers is a *two-variable exponential equation.* Like all two-variable equations, the solution set is a set of ordered pairs that make the equation true.

For example, the equation $y = 2^x$ has the ordered pairs (0, 1), (1, 2), (2, 4), and (3, 8) as four of the elements of its solution. Other points can be found by creating a chart where different integer values are chosen for x and the y-value is then calculated.

x	2^x	y
3	2^3	8
2	2^2	4
1	2^1	2
0	2^0	1
−1	$2^{-1} = \dfrac{1}{2^1}$	$\dfrac{1}{2}$
−2	$2^{-2} = \dfrac{1}{2^2}$	$\dfrac{1}{4}$
−3	$2^{-3} = \dfrac{1}{2^3}$	$\dfrac{1}{8}$

When the seven ordered pairs from the chart are graphed, it looks like this. The domain of this graph is all real numbers.

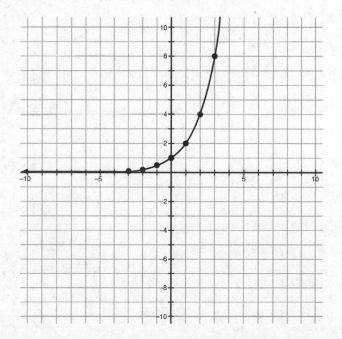

For a two-variable exponential equation with a *b*-value greater than 1, the shape of the graph has this shape, like a playground slide going up to the right.

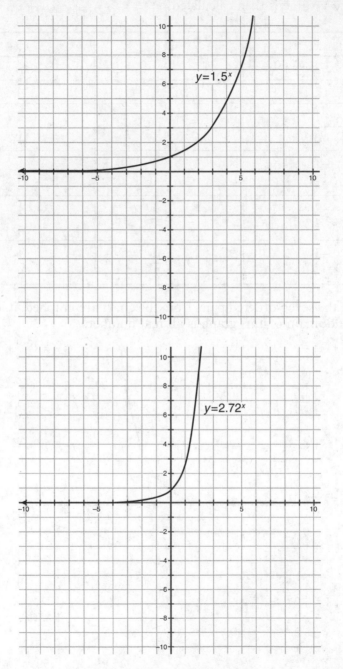

For a two-variable exponential equation with a b value between 0 and 1, the shape of the graph has this shape, like a playground slide going down to the right.

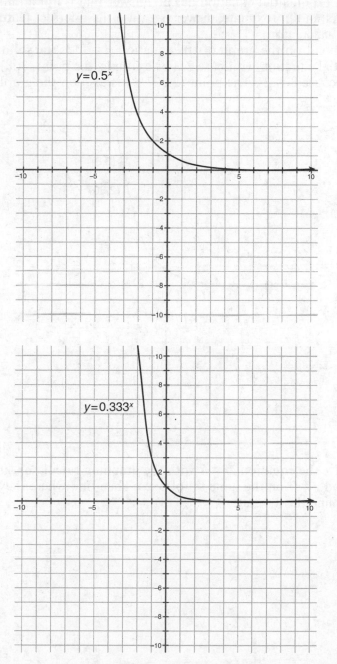

Solving Exponential Equations with a Graphing Calculator

When an exponential equation has an answer that is not an integer, one way to get an approximate answer is to use the intersect feature of the graphing calculator.

Each point on the graph of the equation $y = 2^x$ is the solution to an exponential equation with base 2. For example, the point $(3, 8)$ is on the graph because $8 = 2^3$. The point $(2.807, 7)$ is also on the graph because $7 \approx 2^{2.807}$.

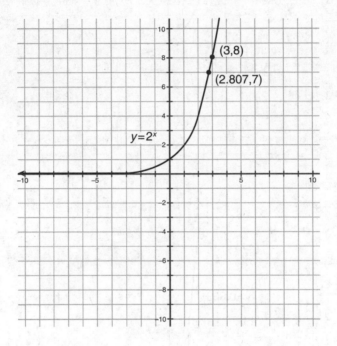

To solve the equation $2^x = 7$ with the graphing calculator, graph both $y = 2^x$ and $y = 7$ on the same set of axes. Then find where they intersect to determine the x-value.

For the TI-84:

Press [y=] and put 2^x after Y1 and 7 after Y2. Press [ZOOM] [6] to see the graphs. Then press [2nd], [TRACE], [5], [ENTER], [ENTER]. Move the cursor near the intersection and press [ENTER] again.

For the TI-Nspire:

From the home screen, press [B] to get to the graphing scratchpad. Then press [tab]. In the entry line, put 2^x after f1(x)= and press [enter] and put 7 after f2(x)= and press [enter]. Press [menu], [6], [4]. Move the cursor to the left of the intersection. Press [enter]. Then move the cursor to the right of the intersection and press [enter] again.

The approximate solution of $2^x = 7$ is shown to be about 2.81.

Example 8

Use the graphing calculator to find, to the nearest hundredth, the solution to the equation $3^x = 5$.

Solution: Using either the TI-84 or the TI-Nspire, the intersection of the graphs of $y = 3^x$ and $y = 5$ is at (1.46, 5).

95

Check Your Understanding of Section 3.2

A. Multiple-Choice

1. Solve $4^x = 256$.
   (1) 4   (2) 8   (3) 16   (4) 64

2. Solve $6^x = \dfrac{1}{36}$.
   (1) 2   (2) −2   (3) −3   (4) −6

3. Which value is nearest to x if $3^x = 40$?
   (1) 13.3   (2) 3   (3) 3.2   (4) 3.4

4. Solve $8^x = 16$.
   (1) $\dfrac{4}{3}$ (2) 2   (3) $\dfrac{3}{4}$ (4) $\dfrac{3}{2}$

5. Solve $7^x = 1$.
   (1) −1   (2) 1   (3) 0   (4) $\dfrac{1}{7}$

6. Solve $9^x = \dfrac{1}{81}$.
   (1) $\dfrac{1}{2}$ (2) −2   (3) 2   (4) $-\dfrac{1}{2}$

7. Solve $4^{2x+3} = 4^{5x-9}$.
   (1) 1   (2) 2   (3) 3   (4) 4

8. Solve $8^{x+4} = 2^{x+4}$.
   (1) 2   (2) −2   (3) 1   (4) −4

9. Which of the following is the graph of $y = 3^x$?

(1)

(2)

(3)

(4)

10. The graph of which equation is shown?

(1) $y = 2^x$ (2) $y = 3^x$ (3) $y = 2^{-x}$ (4) $y = 3^{-x}$

B. Show how you arrived at your answers.

1. Below is a graph of $y = 20$ and $y = 3^x$. Based on the graph, what is an approximate solution to the equation $3^x = 20$?

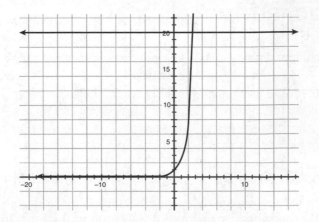

2. Charles says this is the graph of $y = \left(\dfrac{1}{2}\right)^x$. Hope says that it is the graph of $y = 2^{-x}$. Who is correct?

3. Solve for x if $\left(\dfrac{1}{4}\right)^{x-3} = \left(\dfrac{1}{8}\right)^{2x-6}$.

4. $4^x = 100$ is between which two integers? Explain your reasoning?

5. Bianca is solving the equation $3 \cdot 2^x = 36$ as follows:

$$3 \cdot 2^x = 36$$
$$6^x = 36$$
$$x = 2$$

Find the error in Bianca's reasoning.

3.3 LOGARITHMS

KEY IDEAS

An alternative to using graphs to solve an exponential equation is to use a *logarithm*. The logarithm feature of a calculator can be used to solve exponential equations quickly and accurately.

Using the Definition of Logarithm to Solve Equations

A *logarithm* is the number that a number needs to be raised to in order to get another number. Logarithms are written in the form $\log_b a$, where

b is known as the base of the logarithm. To evaluate the expression $\log_2 32$, ask yourself, "To what power must 2 be raised to get 32?" Since $2^5 = 32$, the value of $\log_2 32$ is 5.

Example 1

What is the value of $\log_5 125$?

Solution: This is the same as asking $5^x = 125$. Since $5^3 = 125$, $\log_5 125 = 3$.

Example 2

What is the value of $\log_4 \dfrac{1}{64}$?

Solution: When a positive number is raised to a negative power, the result is a number less than 1. In this case since $4^{-3} = \dfrac{1}{4^3} = \dfrac{1}{64}$, the answer is -3.

Example 3

What is the value of $\log_7 1$?

Solution: Since $7^0 = 1$, $\log_7 1$ must be equal to 0.

MATH FACTS

If the log has no base written, it is assumed to be base 10. So log 100 is the same as the expression $\log_{10} 100$, which is equal to 2 because $10^2 = 100$.

Converting Log Equations into Equivalent Exponential Equations

Any log equation can be converted into an equivalent exponential equation. The equation $\log_b a = c$ can be rearranged to the equation $b^c = a$.

Example 4

Rewrite the equation $\log_5 9 = x$ as an equation that does not involve logarithms.

Solution: $5^x = 9$

Example 5

Rewrite the equation $3^x = 17$ as an equation that involves logarithms.

Solution: $\log_3 17 = x$

Calculating Logarithms with Base 10 or Base *e* with a Calculator

All scientific and graphing calculators have two buttons for logarithms. The log button is for log base 10, and the ln button is for log base *e*. For other bases, the two calculators used in this book have a built-in function for calculating logarithms.

The log button is used to evaluate expressions like $\log_{10} 50$. The expression $\log_{10} 50$ can also be written with the shorthand $\log 50$. Since $10^1 = 10$ and $10^2 = 100$, we should expect $\log_{10} 50$ to be something between 1 and 2. By using the log key on the TI-84 or by pressing [ctrl] [10^x] on the TI-Nspire, we see that $\log_{10} 50 \approx 1.69897$.

The number *e*, like the number π, is an irrational number. It equals approximately 2.72, although a better approximation is 2.71828182845904523536. The number *e* serves an important role in exponential equations used in real-world modeling.

The ln button is used to evaluate expressions like $\log_e 50$. The expression $\log_e 50$ can also be written with the shorthand $\ln 50$. Since $e^3 = 2.72^3 = 7.39$ and $e^4 = 2.72^4 = 54.6$, we should expect $\log_e 50$ to be something between 3 and 4. By using the ln key on the graphing calculator, we see that $\log_e 50 \approx 3.9120$.

MATH FACTS

Using log x is a shorthand way of writing $\log_{10} x$. Using ln x is a shorthand way of writing $\log_e x$. The number e is an important mathematical constant that is equal to approximately 2.72. It is **not** a variable.

Example 6

Evaluate $\log_{10} 750$ to the nearest hundredth.

Solution: Typing log 750 into the calculator gives approximately 2.88.

Example 7

Evaluate $\log_e 750$ to the nearest hundredth.

Solution: Typing ln 750 into the calculator gives approximately 6.62.

Calculating Logarithms with Bases Other than 10 or e

The expression $\log_5 50$ is somewhere between 2 and 3 because $5^2 = 25 < 50$ while $5^3 = 125 > 50$. What is the answer exactly? Since it is an irrational number, getting a decimal approximation to the nearest thousandth is generally all we need.

The TI-84 and the TI-Nspire both have functions that calculate these logarithms.

For the TI-84:
Press [MATH] and then scroll to menu option A for logBASE. Enter 5 as the small number after the log and 50 as the large number.

For the TI-Nspire:
Press the log key by pressing [ctrl] and then [10^x]. Fill in the number 5 as the base and 50 as the number.

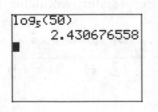

So $\log_5 50 \approx 2.43068$.

If your calculator does not have the log base function, another way to calculate $\log_b a$ is by calculating $\dfrac{\log a}{\log b}$. For $\log_5 50$, this would be $\dfrac{\log 50}{\log 5}$. If you use this method, make sure the numerator and the denominator both use the same type of logarithm (log or ln) but not both. So $\dfrac{\log 50}{\log 5} = \dfrac{\ln 50}{\ln 5}$. However, you cannot use $\dfrac{\log 50}{\ln 5}$ or $\dfrac{\ln 50}{\log 5}$.

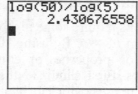

Example 8

Estimate $\log_4 200$ to the nearest hundredth.

Solution: Use the log base function on your calculator: 3.82

Example 9

Estimate $\log_2 0.35$ to the nearest hundredth.

Solution: Use the log base function on your calculator: -1.51

Solving Exponential Equations with a Calculator

Section 3.2 showed you how to solve exponential equations like $2^x = 7$ by graphing $y = 2^x$ and $y = 7$ and then finding the x-coordinate of the intersection point. Now you can solve the same equation by converting the equation into $\log_2 7 = x$ and using the calculator to evaluate $\log_2 7$.

$$2^x = 7$$
$$x = \log_2 7$$
$$x \approx 2.8074$$

Example 10

Estimate the solution to the equation $4^x = 450$. Round to the nearest hundredth.

   Solution:

$$4^x = 450$$
$$x = \log_4 450$$
$$x \approx 4.41$$

Multistep Exponential Equations

Some exponential equations require some algebra steps to convert them into equations in the form $b^x = a$. By using the addition, subtraction, multiplication, and division properties of equality, this can often be accomplished in two steps. First eliminate the constant by adding or subtracting. Then eliminate the coefficient by multiplying or dividing.

   For example, the equation $2 \cdot 3^x + 5 = 50$ could be solved in this way.

$$2 \cdot 3^x + 5 = 50$$
$$\underline{-5 = -5}$$
$$\frac{2 \cdot 3^x}{2} = \frac{45}{2}$$
$$3^x = 22.5$$
$$x = \log_3 22.5$$
$$x \approx 2.834$$

Sometimes the exponent is not just an x but, instead, is a more complicated expression. When this happens, there is more algebra to do after the log step to isolate the x.

   For example, the equation $7^{2x-3} = 19$ could be solved in this way.

$$7^{2x-3} = 19$$
$$2x - 3 = \log_7 19$$
$$2x - 3 \approx 1.513$$
$$\underline{+3 = +3}$$
$$2x \approx 4.513$$
$$x \approx 2.257$$

Example 11

Solve for x rounded to the nearest hundredth in the following equation $3 \cdot 5^{5x-2} + 2 = 200$.

Solution:

$$3 \cdot 5^{5x-3} + 2 = 200$$
$$\underline{-2 = -2}$$
$$\frac{3 \cdot 5^{5x-3}}{3} = \frac{198}{3}$$
$$5^{5x-3} = 66$$
$$5x - 3 = \log_5 66$$
$$5x - 3 \approx 2.603$$
$$\underline{+3 = +3}$$
$$\frac{5x}{5} \approx \frac{5.603}{5}$$
$$x \approx 1.12$$

Leaving the Solution to an Exponential Equation in Unsimplified Form

Sometimes in a multiple-choice question about an exponential equation, the answer choices are not simply numbers. Instead, they are more involved mathematical expressions involving logarithms. These can be solved without using the calculator log functions.

Example 12

Which expression is a solution to the equation $3e^{5x} = 300$?

(1) $\dfrac{\ln 100}{5}$ (2) $\dfrac{\ln 300}{5}$ (3) $\dfrac{\ln 5}{300}$ (4) $\dfrac{\ln 5}{100}$

Solution: Choice (1) is correct.

$$\frac{3e^{5x}}{3} = \frac{300}{3}$$
$$e^{5x} = 100$$
$$5x = \log_e 100$$
$$\frac{5x}{5} = \frac{\log_e 100}{5}$$
$$x = \frac{\log_e 100}{5} = \frac{\ln 100}{5}$$

It is also possible to answer this question by calculating the solution to the original equation, which is $x \approx 0.921$. Then check each of the answer choices to see which one is also approximately equal to 0.921.

The Graph of a Two-Variable Logarithmic Equation

The graph of the function $y = \log_2 x$ is closely related to the graph of the function $y = 2^x$. The ordered pairs (1, 2), (2, 4), (3, 8), (0, 1), and $\left(-1, \frac{1}{2}\right)$ are solutions to $y = 2^x$. The ordered pairs (2, 1), (4, 2), (8, 3), (1, 0), and $\left(\frac{1}{2}, -1\right)$ are solutions to $y = \log_2 x$. Notice that for each ordered pair that satisfies $y = 2^x$, there is a "partner" ordered pair that satisfies $y = \log_2 x$ in which the x-coordinate and y-coordinate have been "swapped." When both equations are graphed on the same set of axes, the graph of $y = \log_2 x$ is the reflection of $y = 2^x$ over the line $y = x$.

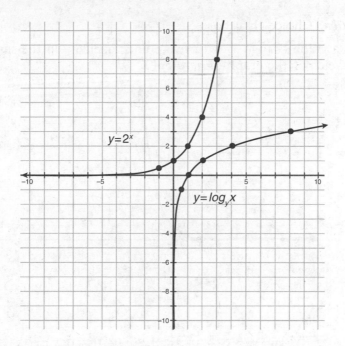

Note that $y = 2^x$ has points only in quadrants I and II and has a horizontal asymptote at $y = 0$. Also note that $y = \log_2 x$ has points only in quadrants I and IV and has a vertical asymptote at $x = 0$. The domain of $y = \log_2 x$ is $x > 0$.

Logarithmic graphs can also be created on the graphing calculator.

Example 13

Below is a graph of $y = 3^x$ with five points labeled. On the same axes, make a sketch of the graph of $y = \log_3 x$.

Solution:

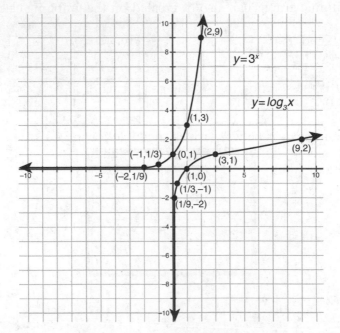

Check Your Understanding of Section 3.3

A. *Multiple-Choice*

1. What is $\log_2 32$?
   (1) 3   (2) 4   (3) 5   (4) 16

2. What is $\log_3 \dfrac{1}{81}$?
   (1) $\dfrac{1}{4}$ (2) $-\dfrac{1}{4}$ (3) 4   (4) -4

3. The equation $m^x = n$ is equivalent to which of the following?
   (1) $\log_n m = x$ (3) $\log_m n = x$
   (2) $\log_m x = n$ (4) $\log_n x = m$

4. $\log_{10} 500$ is between which two integers?
   (1) 2 and 3   (3) 4 and 5
   (2) 3 and 4   (4) 5 and 6

5. Rounded to the nearest tenth, what is the solution to $10^x = 350$?
   (1) 2.3   (2) 2.4   (3) 2.5   (4) 2.6

6. $\log_5 35$ is equal to which of the following?
   (1) $\dfrac{\log 5}{\log 35}$ (2) $\log 7$ (3) $\log 35$ (4) $\dfrac{\log 35}{\log 5}$

7. What is the solution to x to the nearest tenth in $3 \cdot 4^x = 270$?
   (1) 3.0   (2) 3.1   (3) 3.2   (4) 3.3

8. The number e is between which of the following?
   (1) 2 and 3   (3) 4 and 5
   (2) 3 and 4   (4) 5 and 6

9. Which equation has this graph?

(1) $y = \log_4 x$

(3) $y = 4^x$

(2) $y = \log_{\frac{1}{4}} x$

(4) $y = \left(\dfrac{1}{4}\right)^x$

10. If the equation of the graph under the $y = x$ line is $y = \log_{\frac{3}{2}} x$, what is the equation of the graph above the $y = x$ line?

(1) $y = x^{\frac{3}{2}}$ (2) $y = \left(\dfrac{3}{2}\right)^x$ (3) $y = \dfrac{3}{2}x^2$ (4) $y = \left(\dfrac{2}{3}\right)^x$

B. *Show how you arrived at your answers.*

1. Which is greater: $\log_3 1{,}000$ or $\log_4 1{,}000$? Explain your reasoning.

2. If $x = \log_7 7^{123456789}$, what is the value of x?

3. Solve $3 \cdot 5^x + 7 = 262$ for x. Round to the nearest hundredth.

4. Logan solves $7^x = 263$ by calculating $x = \dfrac{\log 263}{\log 7}$. Calvin solves the same equation by calculating $x = \dfrac{\ln 263}{\ln 7}$. Who is correct and why?

5. Make a sketch of $y = \log_2 x$. Include all seven points that have coordinates where one (or both) of the coordinates are integers.

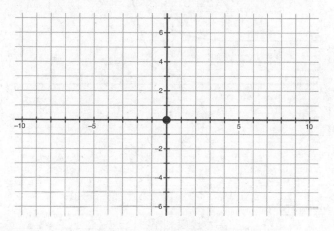

3.4   TRANSFORMED GRAPHS OF EXPONENTIAL AND LOGARITHMIC FUNCTIONS

KEY IDEAS

By knowing the graph of the basic exponential and logarithmic functions like $f(x) = 2^x$ or $g(x) = \log_2 x$, it is possible to create or identify the graphs of more complicated functions involving exponential or logarithmic expressions. Doing so requires you use transformations like horizontal and vertical shifts and also horizontal and vertical stretches and squeezes.

Graphs with Vertical Shifts

The graphs of $y = 2^x$ and $y = 2^x + 3$ are closely related. The graph of $y = 2^x + 3$ is what you get when each point on the graph of $y = 2^x$ is shifted up by 3 units. The horizontal asymptote of $y = 2^x + 3$ is at $y = 3$ since the asymptote of $y = 2^x$, which is $y = 0$, is also shifted up by 3 units.

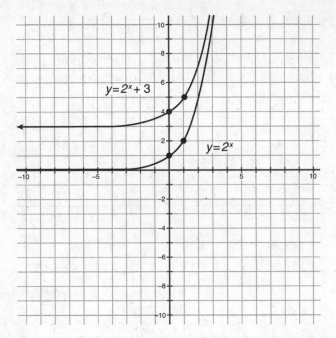

The graphs of $y = \log_2 x$ and $y = \log_2 x + 3$ are related in the same way. The graph of the second equation is the same as the graph of the first but shifted up by 3 units.

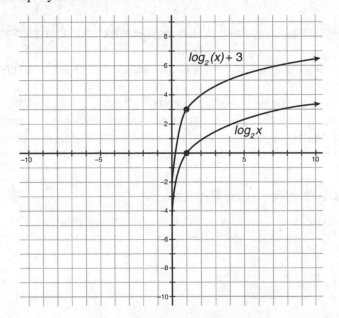

=== **MATH FACTS** ===

In general, the graph of $y = f(x) + a$ is a vertical shift of a units of the graph of $y = f(x)$. If a is positive, the shift is up. If a is negative, the shift is down.

Example 1

If $f(x) = 3^x$, make a sketch of the graphs of $y = f(x)$ and $y = f(x) - 2 = 3^x - 2$ on the same set of axes.

Solution: The graph of $y = f(x) - 2$ is a vertical shift down by 2 units of the graph of $y = f(x)$.

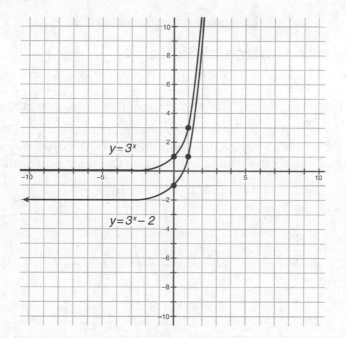

Graphs can also be made using the graphing calculator.

Graphs with Horizontal Shifts

Some of the ordered pairs that satisfy the equation $y = 2^{x+3}$ are $(-2, 2)$, $(-1, 4)$, $(-3, 1)$, $\left(-4, \dfrac{1}{2}\right)$, and $\left(-5, \dfrac{1}{4}\right)$. When these are graphed, they create a curve that is the same as the one created by $y = 2^x$ but shifted 3 units to the left. The horizontal asymptote of both curves is $y = 0$ since a horizontal line shifted to the left remains the same horizontal line.

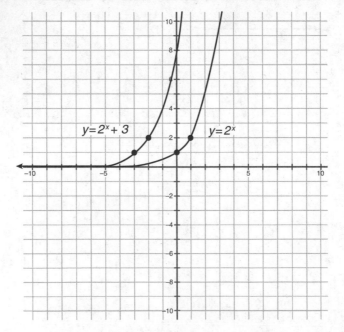

Some of the ordered pairs that satisfy the equation $y = \log_2 (x + 3)$ are $(-1, 1)$, $(1, 2)$, $(-2, 0)$, $\left(-\dfrac{5}{2}, -1\right)$, and $\left(-\dfrac{11}{4}, -2\right)$. When these are graphed, they create a curve that is the same as the one created by $y = \log_2 x$ but shifted 3 units to the left. The vertical asymptote of $y = \log_2 (x + 3)$ is at $x = -3$ but the vertical asymptote of $y = \log_2 x$ is at $x = 0$. Notice that the asymptote has been shifted to the left by 3 units.

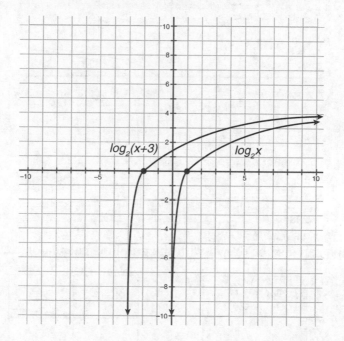

Example 2

If $g(x) = \log_3 x$, sketch on the same set of axes $y = g(x)$ and $y = g(x - 2) = \log_3 (x - 2)$.

Solution: Shifting a curve left by -2 is equivalent to shifting it right by $+2$.

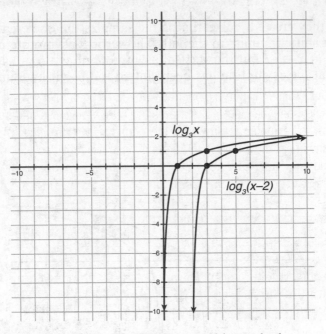

This graph can also be drawn with the graphing calculator.

3.5   USING EXPONENTIAL OR LOGARITHMIC EQUATIONS IN REAL-WORLD SCENARIOS

$$\bigwedge_{\text{KEY IDEAS}}$$

Many real-world scenarios can be modeled with exponential or logarithmic equations. These include exponential growth of populations or of money and also exponential decay of temperature or of radioactive material. If an equation for a model is provided, the equation can be used to answer questions about the scenario. If an equation is not provided, it is possible to create an equation that can then be used to answer questions.

Using an Exponential Equation that Is Already Provided

The population of Regentsland can be modeled by the equation $P = 20 \cdot 1.03^T$, where T is the number of years since 2000 and P is the population in millions. Using this formula, two types of questions can be answered about the model.

Type 1:
Determine the population for a given year.

To find the population in the year 2020, substitute $T = 20$ into the equation and solve for P.

$P = 20 \cdot 1.03^{20} \approx 36$ million people

Type 2:
Determine the year for a given population.

To find when the population of Regentsland will be 50 million people, substitute $P = 50$ into the equation and solve for T. This solution will involve logarithms.

$$50 = 20 \cdot 1.03^T$$
$$\frac{50}{20} = \frac{20 \cdot 1.03^T}{20}$$
$$2.5 = 1.03^T$$
$$T = \log_{1.03} 2.5 \approx 31 \text{ years} = 2031$$

Example 1

The temperature of a hot chocolate can be modeled by the equation $T = 130 \cdot 0.87^M + 70$, where T is the temperature in degrees Fahrenheit and M is the number of minutes since the hot chocolate has been removed from the stove.

a) How hot will the hot chocolate be after 5 minutes?

b) How many minutes will it take for the hot chocolate to be at 86 degrees?

Solution:

a) Substitute 5 for M. $T = 130 \cdot 0.87^5 + 70 \approx 135$ degrees

b) Substitute 86 for M.

$$86 = 130 \cdot 0.87^T + 70$$

$$\frac{-70 = \qquad\qquad -70}{16 = 130 \cdot 0.87^T}$$

$$\frac{16}{130} = \frac{130 \cdot 0.87^T}{130}$$

$$0.1231 = 0.87^T$$

$$M = \log_{.87} 0.1231 \approx 15 \text{ minutes}$$

Example 2

A formula that relates the monthly payment, $\$M$, to pay off a loan of $\$P$ borrowed at an interest rate of r over a period of n months is

$$M = \frac{P\left(\dfrac{r}{12}\right)\left(1 + \dfrac{r}{12}\right)^n}{\left(1 + \dfrac{r}{12}\right)^n - 1}.$$

How many months will it take to pay off a $\$60,000$ student loan at 3% interest if the payment is $\$300$ a month? How much money will have been paid by the time the loan is paid off?

Solution: $P = 60,000$, $r = 0.03$, $M = 300$, and the unknown to solve for is n.

$$300 = \frac{60,000\left(\dfrac{0.03}{12}\right)\left(1 + \dfrac{0.03}{12}\right)^n}{\left(1 + \dfrac{0.03}{12}\right)^n - 1}$$

$$\frac{300}{1} = \frac{150 \cdot 1.0025^n}{1.0025^n - 1}$$

To continue solving this question, cross multiply to create the following equation.

$$300 \cdot 1.0025^n - 300 = 150 \cdot 1.0025^n$$

$$\frac{+300 = \qquad\qquad\qquad +300}{300 \cdot 1.0025^n = 150 \cdot 1.0025^n + 300}$$

The 1.0025^n terms need to be combined to make this into an exponential equation with just one term containing a variable.

$$300 \cdot 1.0025^n = 150 \cdot 1.0025^n + 300$$

$$-150 \cdot 1.0025^n = -150 \cdot 1.0025^n$$

$$150 \cdot 1.0025^n = 300$$

$$\frac{150 \cdot 1.0025^n}{150} = \frac{300}{150}$$

$$1.0025^n = 2$$

$$\log_{1.0025} 2 = n$$

$$277 \approx n$$

It will take approximately 277 months to pay off the loan. The total amount paid in that time will be $300 \cdot 277 \approx \$83,000$.

Creating Exponential Equations for Real-World Scenarios

Many real-world scenarios can be modeled with equations of the form $y = a \cdot (1 + r)^x$, where a and r are replaced with constants. The a-value is the initial value for y, while the r-value is the *growth rate*. When r is positive, the model has exponential growth. When r is negative, the model has exponential decay. The $(1 + r)$ is sometimes called b and is known as the *growth factor* when the equation is in the form of $y = a \cdot b^x$.

Example 3

A viral video has 300 views (V) the first day (D), and the number of views grows at a rate of 15% each day. Create an equation that models this scenario. Use your equation to determine what day the video will have 300,000 views.

Solution: Since 300 is the initial value and 0.15 is the growth rate, the equation is

$$V = 300 \cdot (1 + 0.15)^D$$

To find the day that there are 300,000 views, substitute 300,000 for the variable V and solve using logarithms.

$$300,000 = 300 \cdot 1.15^D$$
$$\frac{300,000}{300} = \frac{300 \cdot 1.15^D}{300}$$
$$1,000 = 1.15^D$$
$$D = \log_{1.15} 1,000$$
$$D = 49 \text{ days}$$

Example 4

A ball is dropped from the top of a 50-foot-tall building. After each bounce, the ball rises to 80% of the highest point of the last bounce. Create an equation that relates the height the ball rises to (H) to the number of bounces (B). Use this equation to determine when the ball will bounce to a height of 10 feet.

Solution: In this situation, the 80% is the growth factor, b, and not the growth rate, r. The equation is $H = 50 \cdot 0.8^B$.

To find how many bounces until the ball bounces to a height of 10 feet, substitute 10 for H in the equation and solve for B.

$$10 = 50 \cdot 0.8^B$$
$$\frac{10}{50} = \frac{50 \cdot 0.8^B}{50}$$
$$0.2 = 0.8^B$$
$$B = \log_{0.8} 0.2 = 7 \text{ bounces}$$

Creating Exponential Equations Based on Compound Interest

Banks generally offer *compound interest*. This means that you get interest on your original money as well as interest on your interest. Different types of interest are compounded annually, compounded monthly, and compounded continuously.

For interest compounded yearly, the formula is $A = P(1 + r)^t$ where P is the initial amount (the principal), r is the interest rate, t is the number of years the money has been in the bank, and A is the amount of money.

If $1,000 is deposited into a bank that offers 3% interest compounded annually, the formula that relates A and P after t years is

$$A = 1,000 \cdot (1 + 0.03)^t = 1,000 \cdot 1.03^t$$

MATH FACTS

For a finite number of compoundings per year, n, the formula that relates the initial amount of money (P) to the amount of money it grows to (A) after a number of years (T) at an interest rate of r is $A = P\left(1 + \dfrac{r}{n}\right)^{nt}$.

Example 5

How long will it take for $1,000 to grow to $2,000 in a bank that offers 1.3% interest compounded monthly?

Solution: Since the interest is compounded monthly, the value of n is 12. The formula is

$$A = 1,000 \cdot \left(1 + \frac{0.013}{12}\right)^{12t} \approx 1,000 \cdot 1.00108^{12t}$$

Substitute 2,000 for A and solve for t.

$$2,000 \approx 1,000 \cdot 1.00108^{12t}$$

$$\frac{2,000}{1,000} \approx \frac{1000 \cdot 1.00108^{12t}}{1,000}$$

$$2 \approx 1.00108^{12t}$$

$$12t \approx \log_{1.00108} 2 \approx 642$$

$$\frac{12t}{12} \approx \frac{642}{12}$$

$$t \approx 54 \text{ years}$$

When the interest is *compounded continuously*, meaning that it is compounded every instant of every second, the formula that relates P and A after t years is $A = Pe^{rt}$. Since the base of the exponential part of the equation is e, the ln button of the calculator can be used to solve for t when A, P, and r are known.

Example 6

How long will it take for $700 to grow to $2,100 if a bank offers 4% interest compounded continuously?

Solution: The equation is $2,100 = 700e^{0.04t}$.

$$2,100 = 700e^{0.04t}$$
$$\frac{2,100}{700} = \frac{700e^{0.04t}}{700}$$
$$3 = e^{0.04t}$$
$$0.04t = \log_e 3 = \ln 3$$
$$0.04t \approx 1.0986$$
$$\frac{0.04t}{0.04} \approx \frac{1.0986}{0.04}$$
$$t \approx 27 \text{ years}$$

Creating Equivalent Exponential Expressions Related to Real-World Scenarios

Exponential equations that arise in real-world scenarios often have a coefficient in front of the exponent. For example, in the equation $A = p \cdot 1.00108^{12t}$, the exponent has a coefficient of 12. It is possible to create an equation that is equivalent to the original equation that has a different coefficient in the exponent. This process utilizes the properties of exponents that $(x^a)^b = x^{ab}$ and that $x^{ab} = (x^a)^b$.

For the example $A = P \cdot 1.00108^{12t}$, the 1.00108^{12t} can be changed to $(1.00108^{12})^t$. This can then be simplified to 1.013^t. So the equivalent equation is $A = P \cdot 1.013^t$.

This process can also work in reverse. If the equation for a scenario involving annual compound interest is $A = P \cdot 1.07^t$, it can be changed into the equivalent expression $A = P \cdot \left(1.07^{\frac{1}{12}}\right)^{12t} \approx P \cdot 1.0057^{12t}$ to make it look more like an equation that represents interest that is compounded monthly. This equivalent expression enables us to conclude that the monthly interest rate is approximately 0.57%.

Check Your Understanding of Section 3.5

A. Multiple-Choice

1. The population of West Algebra can be modeled by the equation $P = 30 \cdot 1.04^T$, where T is the number of years since 2000 and P is the population in millions. How many million people will there be in 2020?
   (1) 63.7   (2) 64.7   (3) 65.7   (4) 66.7

2. The population of Barrontopia can be modeled by the equation $P = 20 \cdot 1.03^T$, where T is the number of years since 2000 and P is the population in millions. In what year will the population be 33 million?
   (1) 2017   (2) 2018   (3) 2019   (4) 2020

3. The temperature of a cup of herbal tea can be modeled by the equation $T = 90 \cdot 0.7^M + 75$, where T is the temperature and M is the number of minutes since the tea was taken off the stove. How hot will the tea be 15 minutes after it is taken off the stove?
   (1) 75 degrees   (3) 79 degrees
   (2) 77 degrees   (4) 81 degrees

4. The temperature of a slice of pizza can be modeled by the equation $T = 70 \cdot 0.85^M + 80$, where T is the temperature and M is the number of minutes the pizza has been out of the oven. When will the pizza be 88 degrees?
   (1) 7 minutes   (3) 11 minutes
   (2) 9 minutes   (4) 13 minutes

5. The amount of money Aria has in the bank after T years is determined by the equation $A = 1,000 \cdot 1.0512^T$. After how many years will Aria have \$2,000 in the bank?
   (1) 12.9   (2) 13.9   (3) 14.9   (4) 15.9

6. The amount of money Zachary has in the bank after T years is determined by the equation $A = 1,000e^{0.05T}$. After how many years will Zachary have \$2,000 in the bank?
   (1) 11.9   (2) 12.9   (3) 13.9   (4) 14.9

7. Which equation relates the population (P) in millions to the time that has passed (T) if the growth rate is 4% per year and the starting population is 10 million people?

(1) $P = 10(0.96)^T$ (3) $P = 10 + 10(0.04)^T$
(2) $P = 10(1.04)^T$ (4) $P = 10(1.4)^T$

8. Which equation relates the amount of money (A) to the amount originally deposited (P) in a bank that offers 2% interest compounded continuously for T years?

(1) $A = Pe^{0.2T}$ (3) $A = P(1 + 0.2)^T$
(2) $A = P(1 + 0.02)^T$ (4) $A = Pe^{0.02T}$

9. Which equation is equivalent to the equation $y = 50(1.07)^{12x}$?

(1) $y = 50(2.1)^x$ (3) $y = 50(1.3)^{6x}$
(2) $y = 50(1.5)^{2x}$ (4) $y = 53.5^{12x}$

10. What is the growth rate for the exponential equation $y = 300(1.7)^x$?

(1) 70% (2) 7% (3) 170% (4) 17%

B. *Show how you arrived at your answers.*

1. The population of the Commonwealth of Common Core Land can be modeled by the equation $P = 8 \cdot 1.07^T$, where P is the population in millions and T is the number of years since 2000.
a) To the nearest million, how big will the population be in 2030?
b) To the nearest year, when will the population be 100 million?

2. The temperature of a pint of ice cream is related to the number of minutes the ice cream has been out of the freezer can be represented by the equation $T = -45 \cdot 0.9^M + 70$.
a) What will the temperature of the ice cream be 12 minutes after it has been taken out of the freezer?
b) How long will it take for the ice cream to reach 41 degrees?

3. Rodney puts $200 into a savings account that offers 12% interest compounded monthly. Create an equation that relates the amount of money in the bank (A) to the number of years it is in the bank (T).

4. Reagan puts $100 into First United North Bank, which offers 5% interest compounded annually. David puts $100 into First United South Bank, which offers 4.9% interest compounded continuously.

Who will have more money in the bank after 2 years?

5. A sheet of paper is $\dfrac{1}{200}$ inches (or $\dfrac{1}{1,267,200}$ miles!) thick. If the paper is folded, its thickness doubles. After n folds, the thickness is

$$T = \dfrac{1}{1,267,200} \cdot 2^N, \text{ where } T \text{ is in miles.}$$

How many folds will it take until the paper's thickness reaches the moon, which is 240,000 miles away?

RADICAL EXPRESSIONS AND EQUATIONS

4.1 SIMPLIFYING RADICALS

∧
KEY IDEAS
⧸___⧹

A *radical expression* is one that involves a $\sqrt{}$ sign (radial sign). Radical expressions are often involved in the solutions to polynomial equations. To work with radicals, you must know how to put them into simplified form and how to combine them.

Definition of Radicals

The square root of a number is the thing that must be multiplied by itself to get that number. The symbol for square root is $\sqrt{}$, also called *the radical sign*. An example is $\sqrt{25} = 5$ because $5^2 = 25$. Even though it is also true that $(-5)^2 = 25$, the symbol $\sqrt{}$ means just the positive number that when squared is equal to the number inside the radical sign. A number that is not a perfect square, like 7, still has a square root although that square root is an irrational number. $\sqrt{7}$ is between 2 and 3 since $2^2 = 4$ and $3^2 = 9$. More precisely, $\sqrt{7}$ is approximately 2.645751311. You can verify this by multiplying 2.645751311 by 2.645751311 to get approximately 7. $\sqrt{25} \cdot \sqrt{25} = 25$, $\sqrt{7} \cdot \sqrt{7} = 7$, and, in general, for a positive value represented by x, $\sqrt{x} \cdot \sqrt{x} = x$.

If there is a small number outside the radical sign, it no longer indicates a square root. If there is a small 3 outside the radical sign, it becomes a *cube root* sign and is equal to the number that must be cubed to become the number under the radical sign. An example is $\sqrt[3]{64} = 4$ because $4^3 = 64$. The small number outside the radical sign is called the *index*. When there is no index, it is implied to be a 2. So a radical sign with no index is called the *square root sign*.

Multiplying Radicals

Two radical expressions that have the same index can be multiplied by multiplying the numbers inside the radical sign. For example $\sqrt{4} \cdot \sqrt{9} = \sqrt{4 \cdot 9} = \sqrt{36} = 6$. This is easily verified for this example since $2 \cdot 3 = 6$.

In general, $\sqrt{a} \cdot \sqrt{b} = \sqrt{ab}$. This rule also works in reverse for factoring radicals: $\sqrt{ab} = \sqrt{a} \cdot \sqrt{b}$.

Simplifying Square Roots

If the number inside a square root sign has a factor that is a perfect square, the radical can be *simplified*. The $\sqrt{50}$ can be simplified since one of the factors of 50 is 25, which is a perfect square.

$\sqrt{50} = \sqrt{25 \cdot 2} = \sqrt{25} \cdot \sqrt{2} = 5 \cdot \sqrt{2} = 5\sqrt{2}$. The 5 in this expression is not an index but a coefficient in front of the radical sign. The multiplication sign between the coefficient and the radical sign is not necessary.

Example 1

Simplify the expression $\sqrt{18}$.

Solution: Since $18 = 9 \cdot 2$, $\sqrt{18} = \sqrt{9} \cdot \sqrt{2} = 3\sqrt{2}$.

It is not always clear whether or not a large number has a factor that is a perfect square. By factoring the number into its *prime factors*, it is possible to group the matching factors into pairs and use the fact that $\sqrt{x} \cdot \sqrt{x} = x$.

$$\sqrt{2592} = \sqrt{2 \cdot 2 \cdot 2 \cdot 2 \cdot 2 \cdot 3 \cdot 3 \cdot 3 \cdot 3} = \sqrt{2 \cdot 2} \cdot \sqrt{2 \cdot 2} \cdot \sqrt{2} \cdot \sqrt{3 \cdot 3} \cdot \sqrt{3 \cdot 3}$$

$$= 2 \cdot 2 \cdot \sqrt{2} \cdot 3 \cdot 3 = 36\sqrt{2}$$

Radical expressions that involve variables can also be simplified with this approach.

Example 2

Simplify the expression $\sqrt{9x^5 y^8}$.

Solution: This can be factored into

$$\sqrt{3 \cdot 3 \cdot x \cdot x \cdot x \cdot x \cdot x \cdot y \cdot y \cdot y \cdot y \cdot y \cdot y \cdot y \cdot y}.$$

Pair up matching factors when possible.

$$\sqrt{3 \cdot 3} \cdot \sqrt{x \cdot x} \cdot \sqrt{x \cdot x} \cdot x \cdot \sqrt{y \cdot y} \cdot \sqrt{y \cdot y} \cdot \sqrt{y \cdot y} \cdot \sqrt{y \cdot y}$$

$$3 \cdot x \cdot x \cdot \sqrt{x} \cdot y \cdot y \cdot y \cdot y$$

$$3x^2 y^4 \sqrt{x}$$

Adding and Subtracting Radicals

Radicals can be added or subtracted only if they have the same index and the same number inside the radical sign. They are combined the same way that like terms are combined with polynomials. For example, $3\sqrt{2} + 4\sqrt{2} = 7\sqrt{2}$. Subtraction works the same way: $8\sqrt{3} - 6\sqrt{3} = 2\sqrt{3}$.

If two radical expressions have the same index but different numbers inside the radical sign, you cannot immediately add or subtract them. Sometimes after simplifying the expressions, they will have the same number inside the radical sign and can then be added or subtracted.

Example 3

Simplify the terms of $\sqrt{50} + \sqrt{18}$ and combine if possible.

Solution: After simplifying each term, it becomes $5\sqrt{2} + 3\sqrt{2} = 8\sqrt{2}$.

Check Your Understanding of Section 4.1

A. Multiple-Choice

1. What is $\sqrt{71} \cdot \sqrt{71}$?
   (1) $\sqrt{71}$ (2) 71   (3) 5,041   (4) 71^2

2. Which of the following is equivalent to $\sqrt{12}$?
   (1) $3\sqrt{2}$ (2) $2\sqrt{6}$ (3) $6\sqrt{2}$ (4) $2\sqrt{3}$

3. Which of the following is equivalent to $\sqrt{18x^9y^6}$?
   (1) $3x^4y^3\sqrt{2x}$ (3) $3x^3y^3$
   (2) $9x^3y^2\sqrt{y^2}$ (4) $3x^3y^2\sqrt{y^2}$

4. What is $2\sqrt{7} + 6\sqrt{7}$?
   (1) $8\sqrt{14}$ (2) $8\sqrt{7}$ (3) $12\sqrt{7}$ (4) $12\sqrt{14}$

5. What is $\sqrt{12} + \sqrt{75}$?
   (1) $7\sqrt{3}$ (2) $\sqrt{87}$ (3) $7\sqrt{6}$ (4) $3\sqrt{7}$

6. Which of the following is equivalent to $\sqrt[3]{192}$?
   (1) $64\sqrt[3]{3}$ (2) $3\sqrt[3]{64}$ (3) $3\sqrt[3]{4}$ (4) $4\sqrt[3]{3}$

7. Which of the following is equivalent to $\sqrt[3]{27x^7y^{11}}$?

(1) $3x^3y^5\sqrt[3]{3xy}$ (3) $3x^2y^3\sqrt[3]{x^2y^2}$

(2) $3x^2y^3\sqrt[3]{xy^2}$ (4) $3x^2y^3\sqrt[3]{xy}$

8. What is $\sqrt[3]{71} \cdot \sqrt[3]{71} \cdot \sqrt[3]{71}$?

(1) $\sqrt{71}$ (2) $\sqrt[3]{71^2}$ (3) 71   (4) 357,911

9. Simplify $\dfrac{6+\sqrt{28}}{2}$.

(1) $3+\sqrt{14}$ (3) $\sqrt{17}$

(2) $3+\sqrt{7}$ (4) $\sqrt{10}$

10. What is $2\sqrt{3} \cdot 5\sqrt{3}$?

(1) 30   (2) $10\sqrt{3}$ (3) $\sqrt{30}$ (4) $7\sqrt{3}$

B. *Show how you arrived at your answers.*

1. Spencer simplifies $\sqrt{25+144}$ as $\sqrt{25}+\sqrt{144}=5+12=17$. Mia simplifies it as $\sqrt{169}=13$. Who is correct?

2. Daniel calculates $\sqrt{49}+\sqrt{576}=\sqrt{625}=25$. Is this correct? Why or why not?

3. In this right triangle, what is the length of leg $\overline{BC}$ simplified in simplest terms?

129

4. What is $\dfrac{6}{\sqrt{3}} \cdot \dfrac{\sqrt{3}}{\sqrt{3}}$?

5. What is $\left(5 + 2\sqrt{3}\right)\left(5 - 2\sqrt{3}\right)$?

4.2 IMAGINARY AND COMPLEX NUMBERS

KEY IDEAS

An *imaginary* number is a number that, when multiplied by itself, results in a negative number. The most basic imaginary number is $\sqrt{-1}$ since $\left(\sqrt{-1}\right)^2 = \sqrt{-1} \cdot \sqrt{-1} = -1$. The number $\sqrt{-1}$ is abbreviated by the symbol i. Based on this definition, $i^2 = -1$. Though this is represented by a letter, i is not a variable. When imaginary numbers are combined with real numbers, the result is called a *complex number*. Complex numbers can be added, subtracted, and multiplied with rules similar to polynomial adding, subtracting, and multiplying.

Simplifying Imaginary Numbers

The square root of any negative number can be represented as an expression involving an i. To simplify $\sqrt{-9}$, split it into $\sqrt{9 \cdot (-1)} = \sqrt{9} \cdot \sqrt{-1} = 3i$. If the number inside the radical is not a perfect square, like $\sqrt{-7}$, the i will be written to the left of the radical sign, $\sqrt{-7} = \sqrt{-1 \cdot 7} = \sqrt{-1} \cdot \sqrt{7} = i\sqrt{7}$. If the number inside the radical has a factor that is a perfect square, like $\sqrt{-18}$, the i will be written between the coefficient and the radical: $\sqrt{-18} = \sqrt{9 \cdot (-1) \cdot 2} = \sqrt{9} \cdot \sqrt{-1} \cdot \sqrt{2} = 3i\sqrt{2}$.

Adding and Subtracting Imaginary Numbers

To add or subtract two imaginary numbers like $5i + 2i$, they are considered to be like terms. So they can be combined as you would do with variables, $5i + 2i = 7i$. For subtraction, it would be $5i - 2i = 3i$.

Multiplying Imaginary Numbers

Multiplying imaginary numbers like $5i \cdot 2i$ is similar to multiplying expressions like $5x \cdot 2x$. Since $5x \cdot 2x$ would be $10x^2$, $5i \cdot 2i$ is $10i^2$. Since $i^2 = -1$, $10i^2 = 10(-1) = -10$.

Example 1

Multiply $3i\sqrt{2} \cdot 5i\sqrt{3}$.

Solution: $3i\sqrt{2} \cdot 5i\sqrt{3} = 15i^2\sqrt{6} = -15\sqrt{6}$

Powers of *i*

i can be raised to other powers besides 2. When *i* is raised to an integer power, the only possible values are *i*, −1, −*i*, or 1.

Power		Solution
0	i^0	1
1	i^1	*i*
2	i^2	−1
3	$i^3 = i^2 \cdot i = (-1)i$	−*i*
4	$i^4 = i^3 \cdot i = (-i)i = -i^2 = -(-1)$	1

After the power of 3, the powers of *i* cycle between the four values: 1, *i*, −1, and −*i*. *i* raised to any multiple of 4 will be 1.

To raise *i* to a high power, like i^{47}, find the multiple of 4 smaller than 47 that is closest to it and rewrite as $i^{44} \cdot i^3 = 1 \cdot i^3 = i^3 = -i$.

The graphing calculator has a built-in *i*. On the TI-84, you access it by pressing [2ND] [.]. On the TI-Nspire, you press the [pi] button to access the constants menu. Calculating powers of *i* on the TI-84 outputs cryptic numbers. If you see something like −3E-13, this is shorthand for −0.0000000000003, which is approximately 0.

Complex Numbers

When a real number, like 3, is added to an imaginary number, like 2*i*, together the sum is written as 3 + 2*i*. A *complex number* is a number of the form *a* + *bi* where *a* and *b* are real numbers.

131

Adding and Subtracting Complex Numbers

Complex numbers are added by adding the real parts and the imaginary parts separately and writing the answer in $a + bi$ form.

$$(3 + 2i) + (5 - 7i) = 8 - 5i$$

Subtracting complex numbers requires care as you distribute the -1 through the second expression.

$$(3 + 2i) - (5 - 7i) = 3 + 2i - 5 + 7i = -2 + 9i$$

The graphing calculator can also add or subtract complex numbers.

Multiplying Complex Numbers

Complex numbers can be multiplied the way binomials are multiplied by using the FOIL shortcut. Anytime an i^2 is encountered in a solution, it should be converted to a -1.

To multiply $(3 + 2i)(5 - 7i)$ with FOIL:

$$3 \cdot 5 + 3(-7i) + 2i(5) + 2i(-7i) = 15 - 21i + 10i - 14i^2$$
$$= 15 - 21i + 10i + 14 = 29 - 11i.$$

The graphing calculator can also multiply complex numbers.

Example 2

Multiply the numbers $3 - 4i$ and $2 + 5i$ and put the answer into $a + bi$ form.

Solution: $(3 - 4i)(2 + 5i) = 6 + 15i - 8i - 20i^2 = 6 + 15i - 8i + 20$
$= 26 + 7i$

Example 3

Write $(4 + 3yi)(3 - 2i) - (4 - 3yi)(3 - 2i)$ in $a + bi$ form, where y is a real number.

Solution: The short way to solve this is by first noticing that there is a $(3 - 2i)$ in each of the main expressions. This can be factored out to make the entire expression:

$$(3 - 2i)((4 + 3yi) - (4 - 3yi))$$
$$(3 - 2i)(4 + 3yi - 4 + 3yi)$$
$$(3 - 2i)(6yi)$$
$$18yi - 12yi^2$$
$$12y + 18yi$$

Complex Solutions to Quadratic Equations

If in the process of using the quadratic formula a negative number appears inside the radical, the quadratic equation is said to have two complex solutions.

Example 4

Use the quadratic equation to solve $x^2 - 4x + 13 = 0$ with $a = 1$, $b = -4$, and $c = 13$.

Solution: $x = \dfrac{-(-4) \pm \sqrt{(-4)^2 - 4(1)(13)}}{2(1)} = \dfrac{4 \pm \sqrt{16 - 52}}{2} = \dfrac{4 \pm \sqrt{-36}}{2}$

Since there is a negative number inside the radical sign, the solutions will be complex numbers.

$$x = \frac{4 \pm 6i}{2} = 2 \pm 3i$$

There are two complex solutions: $x = 2 + 3i$ or $x = 2 - 3i$.

Graphing Complex Numbers on the Complex Plane

A number like $2 + 3i$ cannot be graphed on the standard number line. Instead, complex numbers are graphed on something called *the complex plane*. The complex plane is like a number line for complex numbers. The

133

complex plane resembles the two-axis coordinate plane. Instead of the axes representing the x-coordinate and the y-coordinate, the axes represent the real part of the complex number and the imaginary part of the complex number.

Below is the complex plane. The number $2 + 3i$ is plotted as a single point in the position $(2, 3)$.

Real numbers like 2 can be thought of as $2 + 0i$ and are graphed on the real axis of the complex plane. Imaginary numbers like $2i$ can be thought of as $0 + 2i$ and are graphed on the imaginary axis of the complex plane.

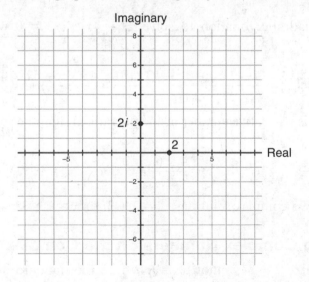

The absolute value of a complex number is the distance on the complex plane between that number and the number $0 + 0i$. The absolute value of a complex number $a + bi$ can be calculated with the formula $|a + bi| = \sqrt{a^2 + b^2}$.

Example 5

What is $|3 - 4i|$?

(1) 5
(2) −5
(3) 7
(4) −7

Solution: $|3 - 4i| = \sqrt{3^2 + (-4)^2} = \sqrt{9 + 16} = \sqrt{25} = 5$ so choice (1) is the answer.

Check Your Understanding of Section 4.2

A. *Multiple-Choice*

1. What is $\sqrt{-25}$?
   (1) $5i$ (2) 5   (3) $-5i$ (4) −5

2. What is $7i - 2i$?
   (1) 5   (2) $5i^2$ (3) −5   (4) $5i$

3. What is $-7i \cdot 3i$?
   (1) 21   (2) −21   (3) $21i$ (4) $-21i$

4. What is i^{66}?
   (1) −1   (2) 1   (3) $-i$ (4) i

5. What is $(3 + 5i) + (2 - 7i)$?
   (1) $5 + 2i$ (3) $-5 - 2i$
   (2) $-5 + 2i$ (4) $5 - 2i$

6. What is $(3 + 5i) - (2 - 7i)$?
   (1) $1 - 2i$ (3) $-1 - 2i$
   (2) $1 + 12i$ (4) $-1 - 12i$

7. What is $(3 + 5i)(2 - 7i)$?
   (1) $-29 - 11i$ (3) $29 - 11i$
   (2) $41 - 11i$ (4) $-41 - 11i$

8. What is $(3 + 5i)(3 - 5i)$?
   (1) 34 (3) -16
   (2) $-16 - 30i$ (4) $34 - 30i$

9. Solve the quadratic equation $x^2 - 6x + 34 = 0$.
   (1) $3 \pm 5i$ (3) $13, -7$
   (2) $-3 \pm 5i$ (4) $-13, 7$

10. What is $|12 - 5i|$?
    (1) $\sqrt{119}$ (2) 12 (3) 13 (4) 7

B. *Show how you arrived at your answers.*

1. Plot the point $-6 + 8i$ on the complex plane. What is the absolute value of $-6 + 8i$?

2. Find the 4 answers to the equation $x^4 + 13x^2 + 36 = 0$.

3. What is $\left(\sqrt{2} + i\sqrt{2}\right)^2$?

4. What is $\left(1 + \dfrac{1}{2}i\right)^2 + \left(1 + \dfrac{1}{2}i\right)$?

5. A complex number whose absolute value is greater than 2 is not in something called the Mandelbrot Set. Show that $1.9 + 0.8i$ is not in the Mandelbrot Set.

4.3 SOLVING RADICAL EQUATIONS

KEY IDEAS

A *radical equation* is one that involves the square root, or cube root, or other kind of root of a variable. If a radical equation also involves the same variable raised to the first power, the equation often has multiple solutions. One step in a radical equation is usually to, at some point, square, or cube, both sides of the equation.

One-Step Radical Equations

When you square the square root of a perfect square, the result is the number that was originally under the radical sign. So $\left(\sqrt{9}\right)^2 = 3^2 = 9$. This is also true for numbers that are not perfect squares like $\left(\sqrt{7}\right)^2 = 7$. In general, we can say $\left(\sqrt{x}\right)^2 = x$.

The equation $\sqrt{x} = 9$ can be solved by squaring the expression on both sides of the equal sign.

$$\sqrt{x} = 9$$
$$\left(\sqrt{x}\right)^2 = 9^2$$
$$x = 81$$

Two-Step Radical Equations

Before squaring both sides of a radical equation, the radical term must be isolated. In the equation $\sqrt{x} + 3 = 7$, the $\sqrt{x}$ must be isolated by subtracting 3 from both sides of the equation. Then the equation can be completed by squaring both sides of the equation to eliminate the radical sign.

$$\sqrt{x} + 3 = 7$$
$$\underline{-3 = -3}$$
$$\sqrt{x} = 4$$
$$\left(\sqrt{x}\right)^2 = 4^2$$
$$x = 16$$

If the expression under the radical sign is not simply an x, there will be additional steps after the radical sign has been eliminated.

$$\sqrt{2x+1} = 5$$
$$\left(\sqrt{2x+1}\right)^2 = 5^2$$
$$2x+1 = 25$$

Now this equation must be solved using techniques from algebra.

$$2x+1 = 25$$
$$\underline{-1 = -1}$$
$$2x = 24$$
$$\frac{2x}{2} = \frac{24}{2}$$
$$x = 12$$

Example 1

Solve for x in the equation $2\sqrt{x+4} - 3 = 11$.

Solution: Two steps are needed to isolate the radical term.

$$2\sqrt{x+4} - 3 = 11$$
$$\underline{\quad +3 = +3}$$
$$2\sqrt{x+4} = 14$$
$$\frac{2\sqrt{x+4}}{2} = \frac{14}{2}$$
$$\sqrt{x+4} = 7$$

Now square both sides, and then isolate the x.

$$\left(\sqrt{x+4}\right)^2 = 7^2$$
$$x+4 = 49$$
$$\underline{-4 = -4}$$
$$x = 45$$

Radical Equations Involving a Linear Term

A more complicated type of radical equation is one like $\sqrt{x} + x = 6$. By testing different values, it can be seen that $x = 4$ is a number that makes the equation true since $\sqrt{4} + 4 = 2 + 4 = 6$. A different approach is needed if the equation requires an algebraic solution, if there is more than one solution, or if the solutions are irrational numbers.

As with the simpler equations, the first step is still to isolate the radical term. For this example, this means to subtract x from both sides of the equation.

$$\sqrt{x} + x = 6$$
$$\underline{-x = -x}$$
$$\sqrt{x} = 6 - x$$

Just like before, the next step is to eliminate the radical sign by squaring both sides of the equation. Squaring the right side, however, involves using polynomial multiplication (FOIL) described in Chapter 1.

$$\left(\sqrt{x}\right)^2 = \left(6 - x\right)^2$$
$$x = 36 - 12x + x^2$$

The problem has been turned into a quadratic equation that can be solved by eliminating all the terms from one side of the equation and then using one of the approaches from Chapter 1 to solve that equation.

$$x = 36 - 12x + x^2$$
$$\underline{-x = -x}$$
$$0 = x^2 - 13x + 36$$
$$0 = (x - 4)(x - 9)$$
$$x - 4 = 0 \quad \text{or} \quad x - 9 = 0$$
$$x = 4 \qquad \text{or} \quad x = 9$$

This process did get the solution $x = 4$, but it also seems to have found another solution. When you substitute $x = 9$ into the original equation, the solution does not work out properly.

$$\sqrt{9} + 9 = 6$$
$$3 + 9 = 6$$
$$12 \neq 6$$

The solution $x = 9$ is not true!

When this happens, the number is not an answer to the equation and we do not include it in the solution set. The clearest way to indicate this on a test is to cross out the $x = 9$ solution and write the word "reject."

The only solution to this equation is $x = 4$.

MATH FACTS

Squaring both sides of an equation can sometimes make equations that are not true into equations that are true. For example, it is not true that $-5 = 5$, but it is true that $(-5)^2 = 5^2$. This is why it is necessary to check any solutions that come from solving a radical equation that eventually became a quadratic equation.

Example 2

Solve for all values of x that satisfy the equation $\sqrt{x} - x = -6$

Solution:

$$\sqrt{x} - x = -6$$
$$\underline{+x = +x}$$
$$\sqrt{x} = -6 + x$$
$$(\sqrt{x})^2 = (-6 + x)^2$$
$$x = 36 - 12x + x^2$$
$$\underline{-x = -x}$$
$$0 = x^2 - 13x + 36$$
$$0 = (x - 4)(x - 9)$$
$$x - 4 = 0 \text{ or } x - 9 = 0$$
$$x = 4 \text{ or } x = 9$$

Each of these potential solutions needs to be substituted back into the original equation to check to see if they work.

Substitute $x = 4$: $\sqrt{4} - 4 = 2 - 4 = -2 \neq -6$. This solution does not work.
Substitute $x = 9$: $\sqrt{9} - 9 = 3 - 9 = -6 = -6$. This solution does work.
The solution is only $x = 9$.

Example 3

Find all solutions to the equation $\sqrt{x-2} + x = 8$.

Solution: Isolate the radical term, square both sides, and solve the resulting quadratic equation.

$$\sqrt{x-2}+x=8$$

$$\frac{-x=-x}{\sqrt{x-2}=8-x}$$

$$\left(\sqrt{x-2}\right)^2=(8-x)^2$$

$$x-2=64-16x+x^2$$

$$\frac{-x+2=-x+2}{0=x^2-17x+66}$$

$$0=(x-11)(x-6)$$

$$x-11=0 \quad \text{or} \quad x-6=0$$

$$x=11 \quad \text{or} \quad x=6$$

Of these two potential solutions, only $x = 6$ works since $\sqrt{11-2}+11=14$, not 8.

Example 4

What is the solution set for the equation $\sqrt{2x}+x-4$?

(1) {2, 8}
(2) {8}
(3) {2}
(4) {}

Solution: Choice (3) is the correct answer.

This question can be solved with algebra like the others and will eventually become a quadratic equation with two solutions, $x = 2$ and $x = 8$. However, the $x = 8$ solution will need to be rejected since it does not make the original equation true.

Since this is a multiple-choice question, the quickest and most accurate way to answer it would be to check the two numbers $x = 2$ and $x = 8$ to see that only $x = 2$ makes the original equation true.

Radical Equations Involving Two Radical Terms

Solving an equation that involves two radical terms like $\sqrt{x+1}+\sqrt{x-4}=5$ is a lengthy process. It is tempting to try to square both sides to eliminate all the radicals. However, this does not work since after squaring the left side with FOIL, there will be outer and inner terms to deal with.

To solve this sort of equation, first isolate one of the radical terms. Then square both sides very carefully. Continue by isolating the other radical term and squaring both sides again.

$$\sqrt{x+1} + \sqrt{x-4} = 5$$

$$\underline{-\sqrt{x-4} = -\sqrt{x-4}}$$

$$\sqrt{x+1} = 5 - \sqrt{x-4}$$

$$\left(\sqrt{x+1}\right)^2 = \left(5 - \sqrt{x-4}\right)^2$$

$$x+1 = 25 - 10\sqrt{x-4} + x - 4$$

$$\underline{-x = -x}$$

$$1 = 21 - 10\sqrt{x-4}$$

$$\underline{-21 = -21}$$

$$-20 = -10\sqrt{x-4}$$

$$\frac{-20}{-10} = \frac{-10\sqrt{x-4}}{-10}$$

$$2 = \sqrt{x-4}$$

$$2^2 = \left(\sqrt{x-4}\right)^2$$

$$4 = x - 4$$

$$\underline{+4 = +4}$$

$$8 = x$$

Check to see if the answer should be rejected.

$$\sqrt{8+1} + \sqrt{8-4} = \sqrt{9} + \sqrt{4} = 3 + 2 = 5$$

The solution is $x = 8$.

Check Your Understanding of Section 4.3

A. Multiple-Choice

1. What is the solution to $\sqrt{x} + 2 = 9$?
 (1) −49 (2) 49 (3) −7 (4) 7

2. What is the solution to $\sqrt{x+2} = 9$?
 (1) −79 (2) 1 (3) 79 (4) −1

3. What is the solution to $3\sqrt{x-2} + 5 = 17$?
 (1) −18 (2) 18 (3) −16 (4) 16

4. What value(s) of x make the equation $\sqrt{x} + x = 20$ true?
  (1) 14   (2) 25   (3) 16   (4) 16, 25

5. What value(s) of x make the equation $\sqrt{x} - x = -20$ true?
  (1) 16   (2) 25   (3) 16, 25   (4) 36

6. Find all solutions to the equation $\sqrt{8x+9} - x = 2$.
  (1) 5, −1   (2) −1   (3) 5   (4) −5, 1

7. Find all solutions to the equation $\sqrt{x+3} + x = 9$.
  (1) 13, 6   (2) 13   (3) 6   (4) 4

8. Find all solutions to the equation $\sqrt{30x+45} - 2x = 3$.
  (1) $-\dfrac{3}{2}, 6$ (2) $-\dfrac{3}{2}$ (3) 6   (4) $\dfrac{3}{2}, -6$

9. For what value(s) does $\sqrt{x} = x$?
  (1) 1   (2) 0   (3) 1, 0, −1   (4) 1, 0

10. Solve for x: $\sqrt{x-3} + \sqrt{x+9} = 6$.
  (1) 6   (2) 7   (3) 8   (4) 9

B. *Show how you arrived at your answers.*

1. Solve for x.

$$5\sqrt{x-4} + 3 = 18$$

2. Solve for x.

$$\sqrt{2x-8} + x = 4$$

3. The equation $x = 3$ has a solution set of $\{3\}$. If you square both sides of the equation, what is the solution set of the new equation?

4. Jace tried to solve the equation $\sqrt{x-4} + \sqrt{x+4} = 4$ by first squaring both sides. He got:

$$\left(\sqrt{x-4} + \sqrt{x+4}\right)^2 = 4^2$$

$$x - 4 + x + 4 = 16$$

$$2x = 16$$

$$x = 8$$

However, 8 is not a solution to the equation. What did Jace do wrong?

5. What value(s) of x satisfy this equation?

$$\sqrt{x+6} - \sqrt{4-x} = 2$$

4.4 GRAPHS OF RADICAL FUNCTIONS

KEY IDEAS

The graphs of equations involving a radical like $y = \sqrt{x}$ or $y = \sqrt[3]{x}$ have shapes that are related to the graphs of the polynomial graphs $y = x^2$ and $y = x^3$, respectively. Graphs of functions involving radicals can be transformed by shifting left, right, up, or down and through vertical stretches and horizontal squeezes.

The Graph of the Square Root Function

The graph of the function $f(x) = \sqrt{x}$ includes the points (0, 0), (1, 1), (4, 2), and (9, 3). Since the $\sqrt{}$ symbol evaluates only to the positive square root of a number, the y-coordinates of all the points (the *range*) are all numbers greater than or equal to 0. Though the concept of imaginary numbers was introduced in the last section for the purpose of graphing on the coordinate axes, there is no place to graph a point like $(-4, 2i)$. So the x-coordinates of the points on this graph (the *domain*) will also be all numbers greater than or equal to 0. For this reason, the entire graph will be in quadrant I.

Below is the graph of the function $f(x) = \sqrt{x}$.

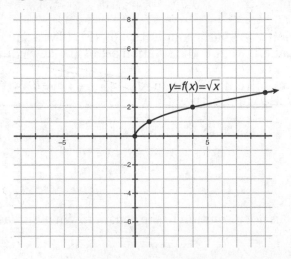

This curve is actually a half of a parabola reflected over the line $y = x$. Unlike the graphs of the polynomial functions introduced in Chapter 1, this graph's domain is not all real numbers but is $x \geq 0$.

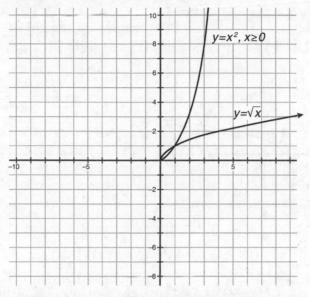

Transformations of the Square Root Function

The function $g(x) = \sqrt{x+2}$ has a graph very similar to the graph of $f(x) = \sqrt{x}$. In fact, $g(x)$ can be written as $f(x + 2)$. The graph of $f(x + 2)$ is the result of shifting the graph of $f(x)$ to the *left* by 2 units.

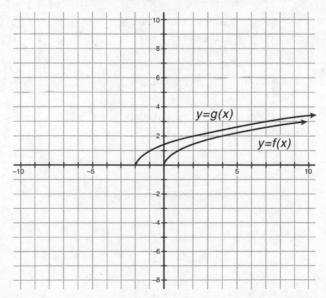

The four most common transformations of the function $f(x)$ are shown in the following table.

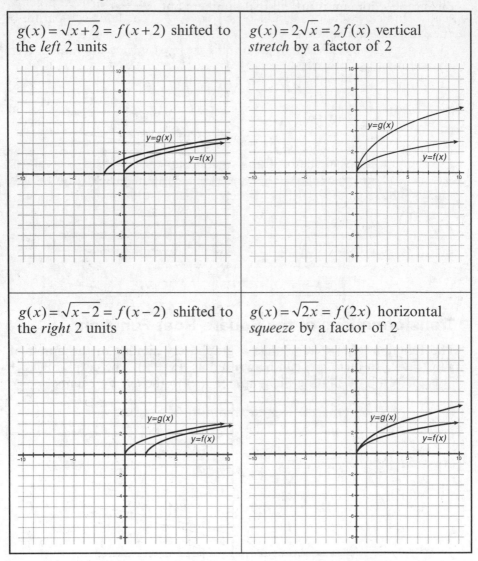

| $g(x)=\sqrt{x+2}=f(x+2)$ shifted to the *left* 2 units | $g(x)=2\sqrt{x}=2f(x)$ vertical *stretch* by a factor of 2 |
| $g(x)=\sqrt{x-2}=f(x-2)$ shifted to the *right* 2 units | $g(x)=\sqrt{2x}=f(2x)$ horizontal *squeeze* by a factor of 2 |

Graphically Solving Radical Equations

In Section 4.3, radical equations like $\sqrt{x} - x = -6$ were solved algebraically. Radical equations can also be solved with the intersect feature of the graphing calculator.

To get the most useful graph to analyze, first isolate the radical term by adding x to both sides.

$$\sqrt{x} - x = -6$$
$$\underline{+x = +x}$$
$$\sqrt{x} = x - 6$$

On the graphing calculator graph on the same set of axes $y = \sqrt{x}$ and $y = x - 6$. Using the intersect feature of the graphing calculator, the two curves intersect at $(9, 3)$. The x-intercept is the solution to the equation $x = 9$.

Check Your Understanding of Section 4.4

A. Multiple-Choice

1. Which graph is a sketch of $y = \sqrt{x} + 2$?

(1)

(2)

(3)

(4)

2. Which graph is a sketch of $y = \sqrt{x+2}$?

(1)

(2)

(3)

(4)

3. Which graph is a sketch of $y = \sqrt{x} - 2$?

(1)

(2)

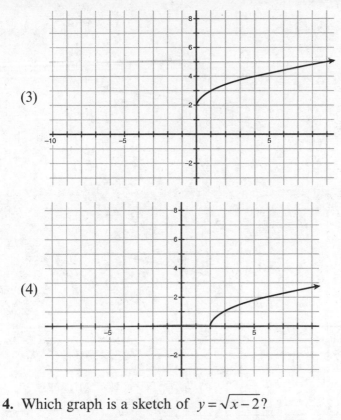

(3)

(4)

4. Which graph is a sketch of $y = \sqrt{x-2}$?

(1)

(2)

(3)

(4)

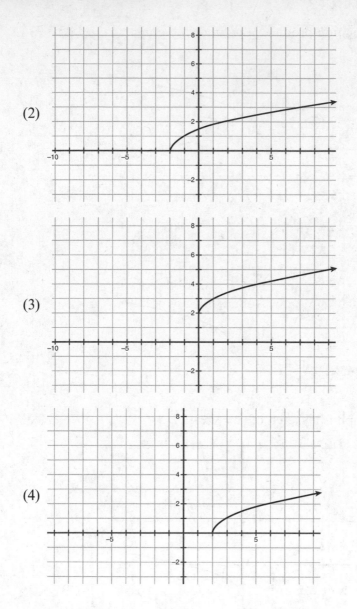

5. Which equation produces the following graph?

(1) $y = \sqrt{x+3} + 1$ (3) $y = \sqrt{x-3} - 1$

(2) $y = \sqrt{x+3} - 1$ (4) $y = \sqrt{x-3} + 1$

6. Which equation produces the following graph?

(1) $y = \sqrt{x-3} + 1$ (3) $y = \sqrt{x-3} - 1$

(2) $y = \sqrt{x+3} + 1$ (4) $y = \sqrt{x+3} - 1$

7. Solve the equation by graphing. $\sqrt{x+3} = -x + 9$
(1) 6   (2) 7   (3) 8   (4) 9

8. Solve the equation by graphing. $\sqrt{8x+9} - x = 2$
(1) (5, –1)   (2) (–5, 1)   (3) 5   (4) –1

9. Solve the equation by graphing. $\sqrt{x-3} + \sqrt{x+9} = 6$
(1) 5   (2) 6   (3) 7   (4) 8

10. Which of the following is a point on the graph of $y = 3\sqrt{2x} + 5$?

(1) (8, 17)   (2) (8, 18)   (3) (8, 19)   (4) (8, 20)

B. *Show how you arrived at your answers.*

1. Sketch the graph of $y = \sqrt{x + 4}$ on the axes below.

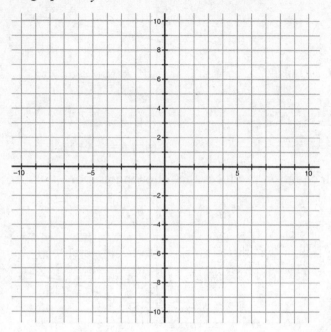

2. What is the equation for this graph?

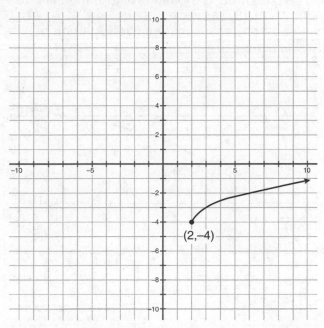

(2,–4)

3. Ariana thinks that this graph is a vertical stretch by a factor of 2 of $y = \sqrt{x}$. Sienna thinks that it is a horizontal squeeze by a factor of 4. Who is correct?

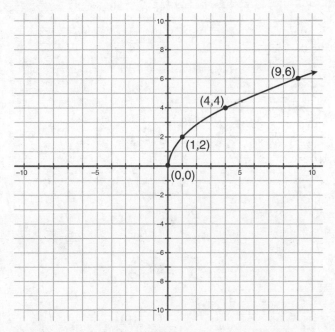

(9,6)

(4,4)

(1,2)

(0,0)

4. How does the graph of $y = \sqrt[3]{x}$ compare to the graph of $y = \sqrt{x}$?

5. Based on this graph, what equation has a solution of $x = 7$?

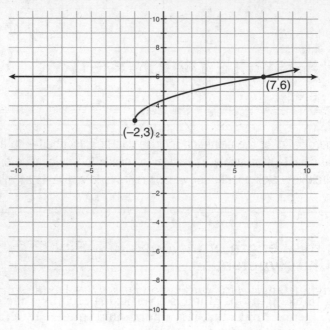

Chapter Five	# TRIGONOMETRIC EXPRESSIONS AND EQUATIONS

5.1   UNIT CIRCLE TRIGONOMETRY

Finding the Length of the Legs of a Right Triangle with a Unit Hypotenuse

Below is right triangle ABC with hypotenuse $AB - 1$ unit. Angle C is the right angle, and angle A has a measure of 40 degrees.

One way to think about sine 40° is that it is the length of the side opposite the 40° angle in a right triangle with a hypotenuse of length 1 unit. If your calculator is in degree mode and you enter sin(40), it should display approximately 0.6428. This means that in triangle ABC, side BC is approximately 0.6428 units long.

If your calculator is not in degree mode, set it to degree mode.

For the TI-84:

Press [MODE], move the cursor to DEGREE on the third line, and press [ENTER] to highlight it. Then press [2nd] [MODE] to get back to calculator mode.

```
NORMAL   SCI   ENG
FLOAT  0123456789
RADIAN  DEGREE
FUNC   PAR  POL  SEQ
CONNECTED  DOT
SEQUENTIAL  SIMUL
REAL  a+bi  re^θi
FULL  HORIZ  G-T
        ↓NEXT↓
```

For the TI-Nspire:

From the home screen, press [5] and then [2] for Document Settings. If the second field does not say "Degree," press [down] and [click] and change the setting to Degree. Then press [tab] until the "Make Default" button is highlighted. Press the [click] button.

If you have a recent operating system on the TI-Nspire, when you are working in the Scratchpad, you will see a small "deg" in the upper right hand corner.

```
Document Settings
   Display Digits: Float 6
          Angle: Degree
Exponential Format: Normal
  Real or Complex: Real
 Calculation Mode: Auto
   Vector Format: Rectangular
 [?] [Restore] [Make Default] [OK] [Cancel]
```

```
Scratchpad ▽                    DEG
```

158

Example 1

In the triangle below, what is the length of side *BC*?

Solution: The length of the side opposite the 72 degree angle in a right triangle with hypotenuse 1 is sine 72°, which is approximately 0.9511.

The cosine of an angle can be thought of as the length of the side adjacent to the angle in a right triangle that has a hypotenuse of length 1 unit.

In triangle *ABC* from above, cosine 40° will display the length of side *AC*, which is adjacent to angle *A*. (Side *AB* also seems to be adjacent to angle *A*. Since *AB* is already the hypotenuse, it can't be two things!) If you enter cos(40) on the calculator, it will display approximately 0.7660.

$\cos(40°) \approx .7660$

=== **MATH FACTS** ===

In a right triangle with hypotenuse of length 1 unit, the length of a side opposite one of the acute angles is the sine of that angle. The length of a side adjacent to one of the acute angles is the cosine of that angle.

Example 2

If AB has length 1 and if angle B has a measure of 53°, find the lengths of sides AC and BC to the nearest hundredth.

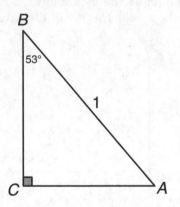

Solution: Since AC is opposite to angle B, its length is sine 53°, which is approximately 0.80. Since BC is adjacent to angle B, its length is cosine 53°, which is approximately 0.60.

If the hypotenuse of a right triangle is known to be 1 and the length of either leg is known, it is possible to find one of the acute angles of the triangle with the $\sin^{-1}$ or the $\cos^{-1}$ function of the calculator.

In triangle ABC, hypotenuse AB has length 1 and leg BC has length 0.6293. Leg BC is opposite $\angle A$. So the measure of $\angle A$ can be found by using the calculator to find which angle has a sine of 0.6293 by using the $\sin^{-1}$ function.

For the TI-84:
Press [2nd], [SIN] to access this feature.
For the TI-Nspire:
Press [trig] and [down] to access this feature.

Since $\sin^{-1}(0.6293) \approx 39°$ (make sure you are in degree mode), the measure of angle A is approximately 39°.

Example 3

In the triangle below, the length of AB is 1 and the length of AC is 0.5446. Rounded to the nearest degree, what is the measure of angle A?

Solution: Since AC is adjacent to angle A, the measure of angle A can be found by calculating $\cos^{-1}(0.5446)$. Rounded to the nearest degree, the angle is 57°.

Even if the hypotenuse of the right triangle is not 1, cosine and sine can be used to find the lengths of the sides if one of the acute angles is known. In the triangle below, the hypotenuse is 5, angle *A* is 40 degrees, and the length of side *BC* is the unknown.

If the hypotenuse were 1, the length of *BC* would be sin 40° ≈ 0.6428. Since the hypotenuse is 5 times greater than 1, the side *BC* will be 5 times greater than 0.6428. Since 5(0.6428) ≈ 3.2140, this is the length of *BC*.

MATH FACTS

If the hypotenuse of the triangle has length *r*, the length of the side opposite angle *A* will be *r* · sin∠*A* and the length of the side adjacent to angle *A* will be *r* · cos∠*A*.

To find an unknown acute angle when one of the legs is known and the hypotenuse is not 1, first divide the lengths of the known sides by the length of the hypotenuse. This changes the triangle into a similar triangle with hypotenuse 1. The angle can then be solved with cos⁻¹ (if the adjacent side was known) or with sin⁻¹ (if the opposite side was known).

Example 4

The length of *AB* is 8 and the length of *BC* is approximately 7.1904. What is the measure of ∠*A* rounded to the nearest degree?

Solution: When the lengths of sides *AB* and *BC* are both divided by 8, which is the length of the hypotenuse, *AB* becomes 1 and *BC* becomes 0.8988. Since *BC* is the side opposite ∠*A*, the measure of ∠*A* can be found with $\sin^{-1}$. Since $\sin^{-1}(0.8988)$ is approximately 64°, this is the measure of ∠*A*.

If in a right triangle only the two legs are known, the Pythagorean theorem can be used to find the hypotenuse first. Then either sine or cosine can be used to find the measure of the angle.

Example 5

The right triangle below has a vertex at (−3, 4). What is the measure of ∠*BOP*? What is the measure of angle *AOP*?

Solution: The length of *OP* can be calculated with the Pythagorean theorem $(-3)^2 + 4^2 = OP^2$. So *OP* = 5. Divide the three sides by 5, and triangle *OPB* becomes:

$$\text{m}\angle BOP = \sin^{-1}\left(\frac{4}{5}\right) \approx 53°$$

Since $\angle AOP$ is supplementary to $\angle BOP$, m$\angle AOP \approx 180° - 53° = 127°$.

Locating Points on the Unit Quarter Circle

The *unit quarter circle* is the part of a circle centered at (0, 0) with a radius of 1 and that is in the first quadrant.

If a radius is drawn so that the angle between radius OP and OA is 27°, it is possible to find the coordinates of the endpoint of the radius P by drawing line segment PR perpendicular to the *x*-axis.

Since triangle ROP is a right triangle with a hypotenuse of 1, the length of the sides can be determined with sine and cosine.
   $OR = \text{cosine } 27° \approx 0.89$
and
   $PR = \text{sine } 27° \approx 0.45$
   The coordinates of point P are (cosine 27°, sine 27°) ≈ (0.89, 0.45).

Example 6

In this unit quarter circle, angle AOP is 61°. What are the coordinates of point P?

   Solution: The coordinates of point P are (cosine 61°, sine 61°) ≈ (0.48, 0.87).

Locating Points on the Full Unit Circle

Even if $\angle AOP$ is greater than 90°, the x-coordinate of point P will be the cosine of $\angle AOP$ and the y-coordinate of point P will be the sine of $\angle AOP$.

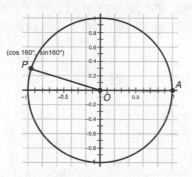

In the unit circle above, $\angle AOP$ is 160°. The coordinates of point P are (cosine 160°, sine 160°). This can be approximated on the calculator as approximately (−0.94, 0.34).

This way of thinking about sine and cosine with a unit circle is related to, but different from, thinking about them with a right triangle.

If you reflect segment OP over the y-axis to become OP', that acute $\angle AOP'$ would be 20°. The acute angle you get when you reflect point P into quadrant I is known as the *reference angle*.

If $\angle AOP$ is greater than 360°, point P "wraps around" and will be in the same location as some other point that is less than 360°. For example, the point at 380° is at the same location as the point 380° − 360° = 20°. This is why the sine of 380° is the same as the sine of 20°.

MATH FACTS

If circle O is a unit circle centered at $(0, 0)$, point A is located at $(1, 0)$, and point P is on the circle, the coordinates of point P are (cosine $\angle AOP$, sine $\angle AOP$). Depending on which quadrant P is in, sine, cosine, or both could be negative.

Example 7

If $\angle AOP$ is between 180° and 270°, what can be inferred about sine $\angle AOP$ and about cosine $\angle AOP$?

(1) They are both positive.
(2) Sine $\angle AOP$ is positive, but cosine $\angle AOP$ is negative.
(3) Sine $\angle AOP$ is negative, but cosine $\angle AOP$ is positive.
(4) They are both negative.

Solution: Choice (4) is correct. Every point in the quadrant III, including point P, has a negative x-coordinate and a negative y-coordinate. Since the x-coordinate is cosine $\angle AOP$ and since the x-coordinate is negative, cosine $\angle AOP$ must also be negative. Since the y-coordinate is sine $\angle AOP$ and since the y-coordinate is negative, sine $\angle AOP$ must also be negative.

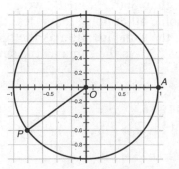

An alternate way to answer this question would be to choose some angle between 180° and 270°, such as 205°. Then enter cos(205) and sin(205) into the calculator (be sure you are in degree mode!) to see that they are both negative.

Example 8

Point P is on unit circle O with point A at $(1, 0)$. If the sine of $\angle AOP$ is negative and the cosine of $\angle AOP$ is positive, what is true about $\angle AOP$?

(1) It is between 0° and 90°.
(2) It is between 90° and 180°.
(3) It is between 180° and 270°.
(4) It is between 270° and 360°.

Solution: On the unit circle, the points that have a positive x-coordinate and a negative y-coordinate are in quadrant IV. The points on the unit circle in quadrant IV range from 270° to 360°, so the answer is choice (4).

Example 9

If sin $\angle AOP$ is negative, in what two quadrants can point P be in?

(1) I or II
(2) II or III
(3) III or IV
(4) I or IV

Solution: Since sine is related to the y-coordinate of point P and since points in quadrant III and IV have negative y-coordinates, the answer is choice (3).

If the coordinates of point P are known, it is possible to determine the measure of $\angle AOP$ with the graphing calculator. Calculating the angle's measure does differ a bit depending on what quadrant point P is in.

If point P is in quadrant I:

$\angle AOP = \sin^{-1}(|y\text{-coordinate}|)$ or $\cos^{-1}(|x\text{-coordinate}|)$

If point P is in quadrant II:

$\angle AOP = 180° - \sin^{-1}(|y\text{-coordinate}|)$ or $180° - \cos^{-1}(|x\text{-coordinate}|)$

If point P is in quadrant III:

$\angle AOP = 180° + \sin^{-1}(|y\text{-coordinate}|)$ or $180° + \cos^{-1}(|x\text{-coordinate}|)$

If point P is in quadrant IV:

$\angle AOP = 360° - \sin^{-1}(|y\text{-coordinate}|)$ or $360° - \cos^{-1}(|x\text{-coordinate}|)$

Example 10

If point P has the coordinates $(-0.8910, -0.4540)$, what is the measure of $\angle AOP$ rounded to the nearest degree?

Solution: A sketch of the unit circle shows that the measure of the angle will be between $180°$ and $270°$.

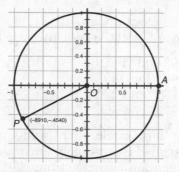

Since point P is in quadrant III, the angle can be calculated with either $180° + \sin^{-1}(0.4540)$ or $180° + \cos^{-1}(0.8910)$. Either way, the answer is $207°$.

Notice that with the process described above, you should always put a positive number into the $\sin^{-1}$ or $\cos^{-1}$ function, even if the x-coordinate or y-coordinate is negative.

Example 11

If $\sin \angle AOP$ is 0.6 and $\cos \angle AOP$ is negative, what is the value of $\cos \angle AOP$?

Solution: Since the sine is positive, point P must be in either quadrants I or II. Since cosine is negative, point P must be in either quadrants II or III. So point P must be in quadrant II since that is where both sine is positive and cosine is negative.

$\angle AOP$ can be calculated with the formula

$180° - \sin^{-1}(|y\text{-coordinate}|) = 180° - \sin^{-1}(0.6) \approx 143°$. Then use the calculator to find cosine $143° \approx -0.8$.

Example 12

If sin $\angle AOP$ is $-\dfrac{5}{13}$ and if cos $\angle AOP$ is positive, what is the value of cos $\angle AOP$?

(1) $\dfrac{12}{13}$

(2) $-\dfrac{12}{13}$

(3) $\dfrac{5}{13}$

(4) $-\dfrac{5}{13}$

Solution: Since the sine is negative, point *P* must be in either quadrants III or IV. Since cosine is positive, point *P* must be in either quadrants I or IV. So point *P* must be in quadrant IV since that is where both sine is negative and cosine is positive.

$\angle AOP$ can be calculated with the formula

$360° - \sin^{-1}(|y\text{-coordinate}|) = 360° - \sin^{-1}\left(\dfrac{5}{13}\right) \approx 337°$. Then use the

calculator to get cosine 337° ≈ 0.9205. Of the choices that are positive, $\dfrac{12}{13} \approx 0.9231$, which is very close to 0.9205. The answer is choice (1).

Example 13

If the coordinates of A are $(-4, 0)$, of O are $(0, 0)$, and of P are $(-3, 4)$, what is the cosine of $\angle AOP$?

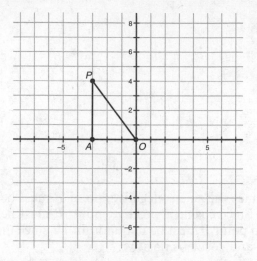

Solution: The sine of $\angle AOP$ is the x-coordinate of the point of intersection (Q) between the unit circle and the line OP. The lengths of the sides of triangle AOP are 3, 4, and 5. Since the hypotenuse OP has a length of 5, dividing all the sides by 5 will create a new triangle with a hypotenuse of 1. The lengths of the sides of similar triangle BOQ are $\dfrac{3}{5}$, $\dfrac{4}{5}$, and 1. Point Q has coordinates $\left(-\dfrac{3}{5}, \dfrac{4}{5}\right)$. So $\cos AOP = -\dfrac{3}{5}$.

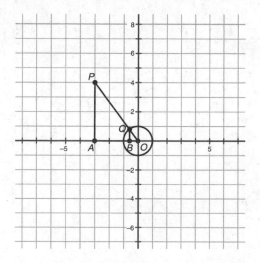

Check Your Understanding of Section 5.1

A. *Multiple-Choice*

1. In right triangle ABC, if $AC = \dfrac{5}{13}$ and $AB = 1$, what is the length of BC?

(1) $\dfrac{5}{12}$ (2) $\dfrac{12}{13}$ (3) $\dfrac{13}{12}$ (4) $\dfrac{12}{5}$

2. In right triangle ABC, the measure of $\angle A$ is 25° and $AB = 1$. What is the length of AC?

(1) 0.9063   (2) 0.4226   (3) 0.2500   (4) 0.2778

3. In right triangle *DEF*, the measure of ∡*D* is 53° and *DE* = 1. What is the length of *EF*?

(1) 0.6018   (2) 0.5889   (3) 0.8000   (4) 0.7986

4. In right triangle *EFG*, if *GE* = 0.7547 and if *FG* = 1, what is the measure of ∡*G*?

(1) 31°   (2) 37°   (3) 41°   (4) 49°

5. In unit circle *O*, if ∡*AOB* = 70°, what are the coordinates of *B*?

(1) (0.94, 0.34)   (3) (0.2, 0.7)
(2) (0.34, 0.94)   (4) (0.4, 0.6)

6. In unit circle O, if $\measuredangle AOB = 110°$, what are the coordinates of B?

(1) (0.94, −0.34)   (3) (−0.41, 0.73)
(2) (−0.2, 0.7)   (4) (−0.34, 0.94)

7. If B is on the unit circle, the x-coordinate of B is positive, and the y-coordinate of B is negative, in which quadrant is point B?

(1) I   (2) II   (3) III   (4) IV

8. In unit circle O, the coordinates of point B are $(-0.91, -0.41)$. What is the measure of $\angle AOB$?

(1) 211° (2) 237° (3) 246° (4) 204°

9. If $\sin \theta > 0°$ and $\cos \theta < 0$ and θ is in the standard position, in which quadrant is the terminal ray of θ?

(1) I (2) II (3) III (4) IV

10. Point B is on unit circle O. What are the coordinates of $\angle \theta$?

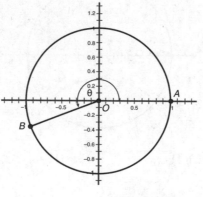

(1) $(-\cos \theta, -\sin \theta)$ (3) $(\cos \theta, -\sin \theta)$
(2) $(-\cos \theta, \sin \theta)$ (4) $(-\sin \theta, \cos \theta)$

B. *Show how you arrived at your answers.*

1. Right triangle *ABC* is similar to triangle *DEF*.

a) What is the length of *AC*?
b) What is the length of *DF*?

2. In right triangle *EFG* with hypotenuse *EG* = 8, what are the lengths of segments *EF* and *FG*?

3. What is the measure of ∡*AOB*?

4. What are three angles between 90° and 360° that have a reference angle of 25°?

5. If $\sin \theta = -0.4848$ and $\theta < 270°$, what is the value of $\cos \theta$?

5.2 TRIGONOMETRY EQUATIONS

A trigonometry equation (often abbreviated as "trig equation") is one where the variable to be solved for is an angle. Trig equations often have multiple solutions, though it is also possible for them to have one or zero solutions.

The equation $\sin x = 0.5$ is an example of a trig equation that has two solutions between 0° and 360°. If you enter sine 30° into a calculator, it will say 0.5. If you, instead, enter sine 150° it will also say 0.5. So the solution set for this equation is {30°, 150°}.

Approximating Solutions to Trig Equations with the Unit Circle

The unit circle explains why there are generally two answers to trig equations. On the axes below is the unit circle and the horizontal line $y = 0.5$. The points on the circumference of the circle are 10 degrees apart from one another. As the horizontal line intersects the circle twice, there are two points on the circle that have a y-coordinate of 0.5, called P_1 and P_2. Also notice that angle BOP_2 is the same measure as angle AOP_1 because P_2 is a reflection over the y-axis of P_1. Angle AOP_1 is a 30° angle, so $\sin 30°$ is 0.5. Angle AOP_2 is a 150° angle, so sine 150° is also 0.5.

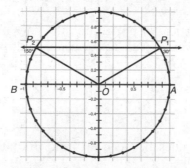

The equation $\sin x = -0.5$ also has two solutions. The line $y = -0.5$ intersects the unit circle at P_1 in quadrant III and at P_2 in quadrant IV. These points correspond to the angles 210° and 330°, respectively.

To use the unit circle to approximate trig equations involving cosine, a vertical line is needed instead.

Example 1

Use the unit circle below to approximate the two solutions (to the nearest 10°) to the equation $\cos x = 0.75$

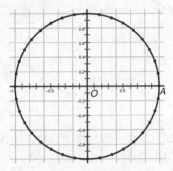

Solution: Points P_1 and P_2 both should have x-coordinates of 0.75 since the x-coordinate of a point on the unit circle is the cosine of the angle. Angle AOP_1 is approximately 40°, while angle AOP_2 is approximately 320°.

In cases where the vertical or horizontal line does not intersect the circle at all, the trig equation is said to have no solution. If the vertical or horizontal line is tangent to the circle, touching it at just one point, the trig equation has just one solution.

Solving Trig Equations More Precisely with a Unit Circle and a Calculator

Finding the two solutions between 0° and 360° to a trig equation like $\sin x = -0.94$ requires four steps.

Step 1:

Sketch the proper vertical or horizontal line on the unit circle. For this example, it is a horizontal line at $y = -0.94$.

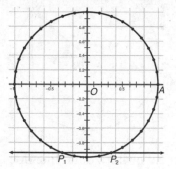

Step 2:

Identify the quadrants for any points of intersection between the unit circle and the line from step 1. For this example, point P_1 is in quadrant III and P_2 is in quadrant IV.

Step 3:

Determine the reference angle by taking the $\sin^{-1}$ or the $\cos^{-1}$ of the absolute value of the number after the equal sign. For this example, the reference angle is $\sin^{-1}(+0.94) \approx 70°$.

Step 4:

Depending on which quadrants the intersection points are in, use the following formulas.

- If point P is in quadrant I, one solution is x = reference angle.
- If point P is in quadrant II, one soution is $x = 180°$ − reference angle.
- If point P is in quadrant III, one solution is $x = 180°$ + reference angle.
- If point P is in quadrant IV, one solution is $x = 360°$ − reference angle.

For this example, since P_1 is in quadrant III and P_2 is in quadrant IV, the two solutions are $x \approx 180° + 70° = 250°$ and $x \approx 360° - 70° = 290°$.

Example 2

Use the unit circle below and the calculator to find the two solutions, to the nearest degree, between $0°$ and $360°$ to the trig equation $\cos x = -0.57$.

Solution: The sketch indicates that there are two points on the unit circle that have x-coordinates of -0.57. P_1 is in quadrant II, and P_2 is in quadrant III.

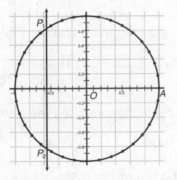

The reference angle is $\cos^{-1}(+0.57) \approx 55°$.

The two solutions are $x \approx 180° - 55° = 125°$ and $x \approx 180° + 55° = 235°$.

Check Your Understanding of Section 5.2

A. *Multiple-Choice*

For questions 1, 2, and 3, use this unit circle diagram.

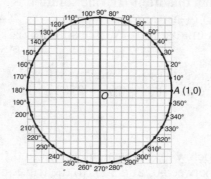

1. Based on the unit circle diagram, find the solution(s) to the equation $\sin x = 0.8$.
   (1) 35°, 325°
   (2) 55°, 305°
   (3) 55°
   (4) 55°, 125°

2. Based on the unit circle diagram, find the solution(s) to the equation $\cos x = -0.6$.
   (1) 55°
   (2) 125°, 235°
   (3) −55°
   (4) 235°, 305°

3. Based on the unit circle diagram, find the solution(s) to the equation $\sin x = -1$.
   (1) 270°   (2) 180°   (3) 90°   (4) 0°

4. Find the solution(s) to $\cos x = 0.5446$.
   (1) 57° only
   (2) 57° and 237°
   (3) 57° and 123°
   (4) 57° and 303°

5. Find the solution(s) to $\cos x = -0.5446$.
   (1) 237° only
   (2) 237° and 303°
   (3) 57° and 303°
   (4) 123° and 237°

6. If $0° \le x < 360°$, what is the maximum number of solutions to the equation $\sin x = a$ where a is a real number?
   (1) 0 solutions   (3) 2 solutions
   (2) 1 solution   (4) 3 solutions

7. Find the solution(s) to $\cos x = -1$ for $0° \le x < 360°$.
   (1) 90°   (2) 0°   (3) 270°   (4) 180°

8. At what value(s) of x does $\cos x = \sin x$?
   (1) 45°   (3) 45° and 225°
   (2) 225°   (4) 45° and 135°

9. If $a°$ is one solution to $\sin x = k$ where $k > 0$, what is another solution?
   (1) $180° - a°$ (3) $360° - a°$
   (2) $180° + a°$ (4) $360° + a°$

10. What is the solution set for $\sin x = -1.5$?
    (1) {215°}   (3) {325°}
    (2) {215°, 325°}   (4) {}

B. *Show how you arrived at your answers.*

1. On this unit circle, draw the two angles that have a sine of 0.7.

2. On this unit circle, draw the two angles that have a cosine of −0.4.

3. How can this unit circle be used to demonstrate that $\sin x = 2$ has no solutions?

4. What are the two solutions to $\sin x = -0.8192$?

5. For which values of x, $0° \leq x < 360°$, is $\sin x > \cos x$?

5.3 RADIAN MEASURE

===== △ KEY IDEAS △ =====

Just as length can be measured in different units like feet or meters, angles can also be measured in different units. In addition to degrees, angles are sometimes measured in a unit called *radians*. One radian is equal to approximately 57.3°.

The circumference of the unit circle is $2 \cdot \pi \cdot 1 = 2\pi \approx 6.28$ units. In the diagram below, the edge of the unit circle is divided into 6 arcs each of length 1 unit and divided into 1 arc of length approximately 0.28 units. The 6 larger angles are each 1 radian. Since there are approximately 6.28 radians in a circle, the conversion factor is that 6.28 radians is approximately 360° making the approximate number of degrees in 1 radian $\dfrac{360°}{6.28} \approx 57.3°$.

The *exact* number of radians in a circle is 2π. So the exact number of degrees in 1 radian is $\dfrac{360°}{2\pi} = \dfrac{180°}{\pi}$.

===== MATH FACTS =====

To convert radian measure to degrees, multiply the number of radians by $\dfrac{180°}{\pi}$.

Example 1

Convert 3.49 radians to degrees, rounded to the nearest degree.

Solution:

$$3.49 \cdot \frac{180°}{\pi} \approx 200°$$

On the TI-84, access the π button by pressing [2ND] [^]. On the TI-Nspire, there is a [pi] button that opens a menu of 10 math symbols, including π.

Example 2

Convert $\dfrac{2\pi}{3}$ radians to degrees. Give an exact answer.

Solution:

$$\frac{2\pi}{3} \cdot \frac{180°}{\pi} = \frac{360°\pi}{3\pi} = 120°$$

MATH FACTS

To convert degrees to radians, divide by $\dfrac{180°}{\pi}$. Since dividing by a fraction can be accomplished by multiplying by the reciprocal of that fraction, it is instead quicker to multiply by $\dfrac{\pi}{180°}$.

Example 3

Convert 50° to radians, rounding to the nearest hundredth of a radian.

Solution:

$$50° \cdot \frac{\pi}{180°} \approx 0.87 \text{ radians}$$

Example 4

Convert 135° to radians. Give an exact answer.

Solution:

$$135° \cdot \frac{\pi}{180°} = \frac{135°\pi}{180°} = \frac{3\pi}{4} \text{ radians}$$

Some common radian to degree conversions are shown in the table.

Degrees	Radians
360°	2π
180°	π
90°	$\dfrac{\pi}{2}$
60°	$\dfrac{\pi}{3}$
45°	$\dfrac{\pi}{4}$
30°	$\dfrac{\pi}{6}$

Angles that are multiples of 30°, 45°, 60°, and 90° can be found by multiplying the radian equivalent of each angle by the appropriate factor.

For example, since 270° = 3 · 90°, in radians $270° = 3 \cdot \dfrac{\pi}{2} = \dfrac{3\pi}{2}$.

Converting Radians and Degrees with a Calculator

The TI-84 and the TI-Nspire both have the ability to convert from degrees to radians or from radians to degrees.

For the TI-84:

To convert radians to degrees, first set the calculator to degree mode. Then enter the number of radians to convert. Put this number in parentheses if it is something involving a fraction. Then press [2ND], [APPS] to get to the ANGLE menu. Press [3] for "r" to convert to radians and press [ENTER].

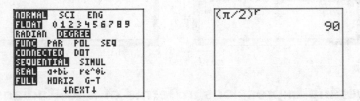

To convert degrees to radians, first set the calculator to radian mode. Enter the number of degrees. Then go to the ANGLE menu. Select degrees by pressing [1] and then press [ENTER].

For the TI-Nspire:

To convert radians to degrees, enter the number of radians. Then press [ctrl], [catalog], and select the radian symbol. Then press [catalog], [1], [D], and select the DD function and press [enter].

To convert degrees to radians, enter the number of degrees. Then press [ctrl], [catalog], and select the degree symbol. Then press [catalog], [1], [R], and select the Rad function and press [enter].

Converting Expressions in Terms of π to Radians

When using the calculator to convert degrees to radians, it will give a decimal approximation of the solution instead of an exact expression involving π. To rewrite as an expression involving π, divide the answer by π. The quotient will be the number that goes in front of the π in the exact answer.

To convert 150° to radians in terms of π on a calculator:

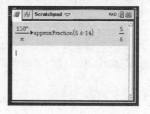

So 150° is equal to $\frac{5}{6}\pi$ radians.

Check Your Understanding of Section 5.3

A. *Multiple-Choice*

1. Convert 30° to radians.

   (1) $\frac{1}{6}$ (2) $\frac{\pi}{4}$ (3) $\frac{\pi}{6}$ (4) $\frac{\pi}{3}$

2. Convert 240° to radians.

   (1) $\frac{4}{3}$ (2) $\frac{5\pi}{6}$ (3) $\frac{4\pi}{3}$ (4) $\frac{5\pi}{4}$

3. Convert $\frac{\pi}{4}$ radians to degrees.

   (1) 90°   (2) 60°   (3) 30°   (4) 45°

4. Convert $\frac{7\pi}{6}$ radians to degrees.

   (1) 200°   (2) 205°   (3) 210°   (4) 215°

5. Convert 41° to radians, rounded to the nearest hundredth of a radian.
   (1) 0.66   (2) 0.68   (3) 0.70   (4) 0.72

6. Convert 308° to radians, rounded to the nearest hundredth of a radian.
   (1) 5.38   (2) 5.40   (3) 5.42   (4) 5.44

7. Convert $\dfrac{1}{6}$ radians to degrees, rounded to the nearest tenth of a degree.

(1) 30.0°   (2) 29.6°   (3) 19.6°   (4) 9.5°

8. Convert 2 radians to degrees, rounded to the nearest tenth of a degree.

(1) 114.6°   (2) 94.6°   (3) 84.6°   (4) 360.0°

9. If circle O has a radius of 1 inch and m∡AOB is 1.3 radians, what is the length of arc $\overset{\frown}{AB}$?

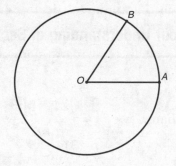

(1) 1 inch   (2) 1.3 inches   (3) 1.6 inches   (4) 1.9 inches

10. In unit circle O, if the length of arc $\overset{\frown}{AB}$ is $\dfrac{\pi}{3}$, what are the coordinates of point B?

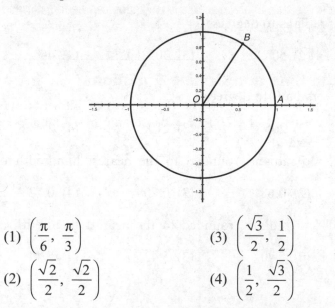

(1) $\left(\dfrac{\pi}{6}, \dfrac{\pi}{3}\right)$ (3) $\left(\dfrac{\sqrt{3}}{2}, \dfrac{1}{2}\right)$

(2) $\left(\dfrac{\sqrt{2}}{2}, \dfrac{\sqrt{2}}{2}\right)$ (4) $\left(\dfrac{1}{2}, \dfrac{\sqrt{3}}{2}\right)$

B. *Show how you arrived at your answers.*

1. Is 1 radian bigger or smaller than 60°? Explain your reasoning.

2. What is $\sin \dfrac{5\pi}{3}$ radians?

3. Arrange from smallest to biggest: 1 radian, 1 right angle, 1 degree.

4. In unit circle O, if $\measuredangle AOB$ is an acute angle measured in radians, what is bigger: $\measuredangle AOB$ or $\sin \measuredangle AOB$?

5. Some calculators in addition to degrees and radians have a "gradians" mode. A gradian is a unit of angle measurement where 400 gradians are in a circle. How many gradians are in $\dfrac{\pi}{2}$ radians?

5.4   GRAPHS OF THE SINE AND COSINE FUNCTIONS

===== KEY IDEAS =====

When a graph is created in which the x-coordinate is the measure of an angle and the y-coordinate is either the sine or the cosine of that angle, the graph has the shape of a wave. This wave shape can be transformed in different ways to make waves that are larger, more compressed, or translated up or down.

Graphs of the Sine and Cosine Functions

Each point on the unit circle conveys three pieces of information: the angle AOP, the cosine of $\angle AOP$ (the x-coordinate of P), and the sine of $\angle AOP$ (the y-coordinate of P).

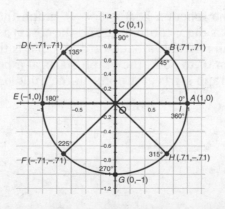

189

The information from this unit circle can be summarized in a chart.

Point	Angle	Cosine(x) (x-coordinate)	Sine(x) (y-coordinate)
A	0°	1	0
B	45°	0.71	0.71
C	90°	0	1
D	135°	−0.71	0.71
E	180°	−1	0
F	225°	−0.71	−0.71
G	270°	0	1
H	315°	0.71	−0.71
I	360° = 0°	1	0

To graph the sine function $f(x) = \sin(x)$, graph the nine points with the angle measure as the x-coordinate and the $\sin(x)$ as the y-coordinate. The final graph will resemble a backward "S" that is tilted on its side.

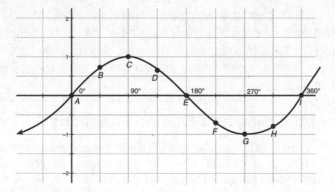

To graph the cosine function $g(x) = \cos(x)$, graph the nine points with the angle measure as the x-coordinate and the $\cos(x)$ as the y-coordinate. The final graph will resemble a backward "S" that is tilted on its side. In a cosine curve from 0° to 360°, the graph starts at a peak. In contrast in a sine curve, the graph starts at a point in the middle.

These graphs can also be created on a graphing calculator.

For the TI-84:

Press [Y=] and enter sin(x) after the Y1=. Press [ZOOM], [7].

For the TI-Nspire:

From the home screen, press [B] to get to the graph scratchpad. If the entry line is not visible, press [tab]. Enter sin(x) after f1(x) = and press [enter]. To get the sine, press the [trig] button. Press [menu], [4], [8] to get a better viewing window.

Since the window is set from −360 to +360, these calculator images show two cycles of the sine curve.

If the angle is in radians, the graph will be the same as when the angle is in degrees. However, the scale on the x-axis will be different. The end of each cycle will be at 2π instead of at 360°. Instead of 0°, 90°, 180°, 270°, and 360°, the angles will be at 0 r, $\frac{\pi}{2}$ r, π r, $\frac{3\pi}{2}$ r, and 2π r. These values correspond to approximately 0 r, 1.57 r, 3.14 r, 4.71 r, and 6.28 r.

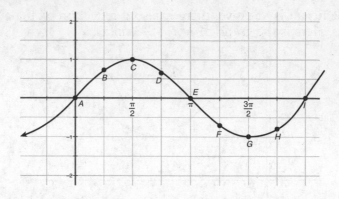

The Amplitude of Sine and Cosine Graphs

In a function like $g(x) = 3\sin(x)$, the number in front of the trig expression is known as the *amplitude*. This coefficient transforms the graph with a *vertical stretch*. The highest point on the graph is now (90°, 3) instead of (90°, 1). The lowest point on the graph is now (270°, −3) instead of (270°, −1).

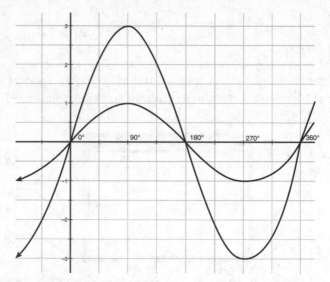

If the coefficient is negative, there is also a reflection of the curve over the x-axis. Here is the graph of $h(x) = -3\sin(x)$.

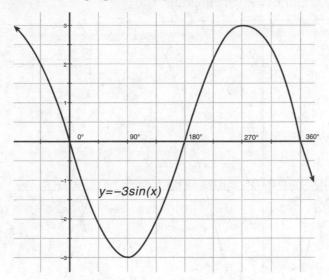

$y=-3sin(x)$

Example 1

Sketch the graph of $f(x) = \cos(x)$ and $g(x) = 2\cos(x)$ on the same set of axes from $0° \le x \le 360°$.

Solution:

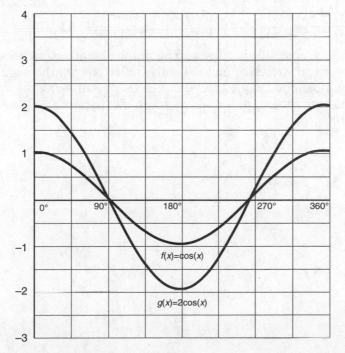

$f(x)=\cos(x)$

$g(x)=2\cos(x)$

Example 2

This is the graph of which equation?

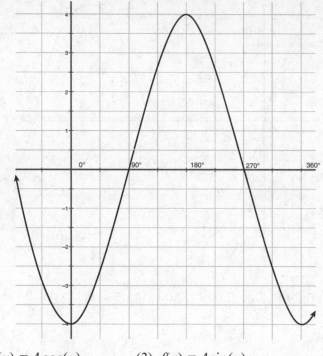

(1) $f(x) = 4\cos(x)$ (3) $f(x) = 4\sin(x)$

(2) $f(x) = -4\cos(x)$ (4) $f(x) = -4\sin(x)$

Solution: Choice (2) is correct. The graph resembles a cosine curve that has a vertical stretch of factor 4 and is reflected over the *y*-axis. Since this is a multiple-choice question, it could also be answered by graphing each of the four choices to see which graph most resembles the graph in the question.

Vertical Translations of Sine and Cosine Graphs

In a function like $g(x) = \sin(x) + 2$, the constant term affects the graph. The graph of $g(x) = \sin(x) + 2$ is like the graph of $f(x) - \sin(x)$ with each point translated up by 2 units. The invisible horizontal line through the middle of the curve, sometimes called the *axis* of the curve, is also translated up 2 units.

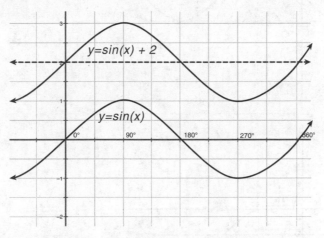

If the constant is negative, the graph will be shifted down.

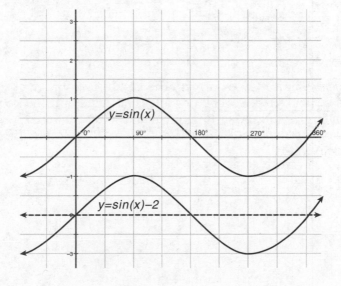

Example 3

Which of the following is the graph of $g(x) = \cos(x) - 1$?

(1)

(2)

(3)

(4)

Solution: Choice (4) is the answer. It is the graph of $f(x) = \cos(x)$ shifted down by 1 unit.

The Frequency of Sine and Cosine Graphs

The graph of the function $g(x) = \sin(2x)$ is like the graph of $f(x) = \sin(x)$ but with a *horizontal squeeze* by a factor of 2. The coefficient of the x, the 2 in this example, is known as the *frequency* since it is the number of complete waves that fit inside a 360° interval.

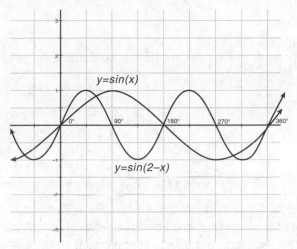

Example 4

Below is the graph of which function?

(1) $g(x) = 3\sin(x)$ (3) $g(x) = \sin(3x)$
(2) $g(x) = \sin(x) + 3$ (4) $g(x) = \cos(3x)$

Solution: Since 3 cycles of the sine curve fit in a 360° interval, the coefficient of the x is 3. Choice (3) is correct.

The Period of Sine and Cosine Graphs

The *period* of a sine or cosine curve is the number of degrees (or radians) needed for the curve to complete one cycle. The period is related to the frequency (B) by the formula $P = \dfrac{360°}{B}$ for degrees or $P = \dfrac{2\pi}{B}$ for radians.

Example 5

What is the period, in degrees, of the curve defined by $g(x) = \cos(4x)$?

Solution: Use the formula.

$$P = \frac{360°}{4} = 90°$$

MATH FACTS

When a function is of the form $g(x) = \pm A\sin(Bx) + C$ or $g(x) = \pm A\cos(Bx) + C$, the A is the amplitude and affects the vertical stretch of the curve. The B is the frequency and affects the horizontal squeeze of the curve. The C affects the vertical translation of the curve. The basic curves $g(x) = \sin(x)$ or $g(x) = \cos(x)$ have values $A = 1$, $B = 1$, and $C = 0$. A function can change one or more of the values to create wavelike curves that pass through different points

---

Graphs of Trig Functions with More than One Transformation

The graph of the function $g(x) = -3\sin(2x) + 1$ is like the graph of $f(x) = \sin(x)$ after four different transformations have been performed on it. The 3 causes a vertical stretch. The negative sign (–) in front of the 3 causes a reflection over the x-axis. The 2 causes a horizontal squeeze. The + 1 causes a vertical translation.

$y=\sin(x)$

$y=-3\sin(2x)+1$

The graphing calculator will graph the transformed function.

Example 6

Which of the following is the graph of $g(x) = 2\cos(3x) + 2$?

Solution: Choice (2) is the answer. It has an amplitude of 2, a frequency of 3, and a vertical shift of 2. It is cosine instead of sine because it starts at a peak instead of in the middle of the curve.

Using the Graphing Calculator to Make a Table of Values

With a graphing calculator, you can quickly display a graph of a complicated trig function on the screen. However, if the question asks you to reproduce the graph accurately on graph paper, a table of values is useful. The table is best if it contains the entries that correspond to the five most important points on the trig graph.

Here are the steps for making a useful table of values for graphing the function $g(x) = -3\sin(2x) + 1$ on paper.

For the TI-84:

Press [Y=] and enter $3\sin(2x) + 1$ after Y1=. Press [2ND] [WINDOW] to open the table setup menu. Set "TblStart" to 0. Since the B-value is 2, set delta Tbl to (360/2)/4. (If you are in radian mode, do (2pi/2)/4 instead). The denominator of the fraction with 360 in the numerator should be the B-value. The independent and dependent settings should both be on "Auto." Press [2nd] [GRAPH] to see the table.

For the TI-Nspire:

In the graph scratchpad, enter $3\sin(2x) + 1$ after f1(x) = in the entry line, and press [enter]. To see the table, hold down [ctrl] and press [T]. Press [menu], [2], [5] to edit the table settings. Make the table start value 0. Since the B-value is 2, set the table step value (360/2)/4 (or for radian mode (2pi/2)/4)) and press the OK button.

Finding the Equation of a Function from a Given Graph

There are five decisions to make when finding the equation of a function on which a trig graph was defined.

1) Is it a sine or a cosine function?

2) What is the vertical shift (the C-value)?

3) What is the amplitude (the A-value)?

4) What is the frequency (the B-value)?

5) Is the coefficient of A positive or negative?

These five questions can be answered in any order.

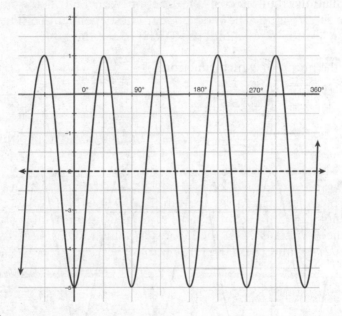

Step 1:

Since the y-intercept of this curve is not in the middle curve but is at one of the high or low points, this is a cosine curve.

Step 2:

Find the value of C, which is the vertical shift. Since the high point is at $y = +1$ and the low point is at $y = -5$, halfway between that is $y = -2$, which is the axis of the curve. So $C = -2$.

Step 3:

Since the distance from the axis to one of the peaks or low points is 3 units, the amplitude is 3. This means $A = 3$.

Step 4:

Since the first cycle of this curve ends at the 90° point, the period is 90°. The formula that relates period (P) and frequency (B) is $P = \dfrac{360°}{B}$. So $90° = \dfrac{360°}{B}$ or $90°B = 360°$. So $B = 4$.

Step 5:

A cosine curve with a negative A-value starts from a low point and goes up to the right.
The equation is $y = -3\cos(4x) - 2$.

Example 7

Below is the graph of which function?

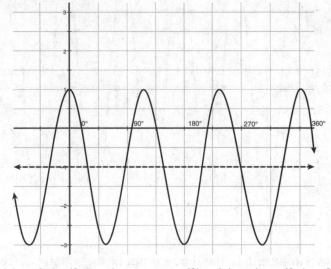

(1) $g(x) = 2\cos(3x) - 1$ (3) $g(x) = 1\cos(3x) - 2$
(2) $g(x) = 3\cos(2x) - 1$ (4) $g(x) = 2\cos(x) - 3$

Solution: Choice (1) is the correct answer. The amplitude (A) is 2. There are 3 (B) curves in the 360° interval. Since the axis is at $y = -1$, the curve has a vertical translation of -1 (C).

Example 8

The angles for this graph are in radians. What is the equation of the function on which this graph is based?

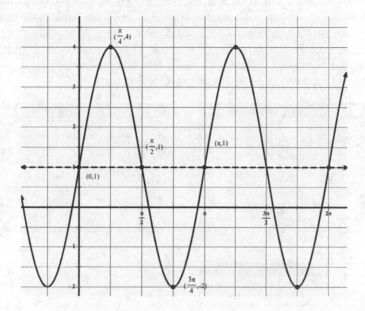

Solution: $f(x) = 3\sin(2x) + 1$. The *C*-value is 1 since the axis is at $y = 1$. The *A*-value is +3 since the high point is 3 units above the axis and the sine curve goes up first. To calculate the *B*-value, solve $P = \dfrac{2\pi}{B}$ with $P = \pi$. So $\pi = \dfrac{2\pi}{B}$, $B\pi = 2\pi$, $B = 2$.

When the angles are in radians, you can use the sinusoidal regression feature of a graphing calculator to find the equation for you.

For the TI-84:

Press [STAT], [1] and enter five points from the graph. The simplest points are the high and low points and also the points that are on the horizontal line that goes through the middle of the curve. The *x*-coordinates go into L1, and the *y*-coordinates go into L2. You can use the [2nd][^] to make a π, and it will turn it into a decimal in the list. Press [STAT], [RIGHT ARROW], and scroll down to [C] for SinReg. Then press [ENTER]. The calculator will tell you values of *a*, *b*, *c*, and *d* for the equation $y = a \cdot \sin(bx + c) + d$.

The $C = -8.80942\text{E-}15$ is -0.0000000000000880942, which is the way the calculator is representing 0.

For the TI-Nspire:

From the home screen, select the Add Lists & Spreadsheet icon. On the spreadsheet, enter x to name the first column and y to name the second column. Then enter the x- and y-values at least five points into columns A and B. Press [menu], [4], and [1] to get to Stat Calculations. Press [C] for Sinusoidal Regression. In the X List box, type x.

In the y List box, type y. Press the OK button.

The equation is $y = 3\sin(2x) + 1$.

Example 9

If the angles are in radians, find the function on which the following graph is based.

Solution: This is a cosine curve with a negative *A*-value. *A* is −10, and *C* is 13. The *B*-value can be calculated by $100 = \dfrac{2\pi}{B}$, $100B = 2\pi$, $B = \dfrac{2\pi}{100} = \dfrac{\pi}{50}$. So the equation is

$$f(x) = -10\cos\left(\frac{\pi}{50}x\right) + 13$$

Check Your Understanding of Section 5.4

A. *Multiple-Choice*

1. What is an equation for this graph?

(1) $y = \cos(x)$ (3) $y = \sin(x)$
(2) $y = -\cos(x)$ (4) $y = -\sin(x)$

2. What is an equation for this graph?

(1) $y = \cos(x)$ (3) $y = \sin(x)$
(2) $y = -\cos(x)$ (4) $y = -\sin(x)$

3. What is an equation for this graph?

(1) $y = \cos(2x)$ (3) $y = \sin(2x)$
(2) $y = 2\cos(x)$ (4) $y = 2\sin(x)$

4. What is the graph of $y = -3\sin(x) + 1$?

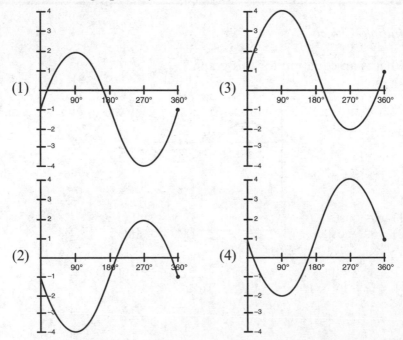

5. What is the y-coordinate of one of the maximum points of $y = 2\cos(x) - 1$?

(1) 3   (2) 2   (3) 1   (4) 0

6. Which of the following is the graph of $y = \sin(2x)$?

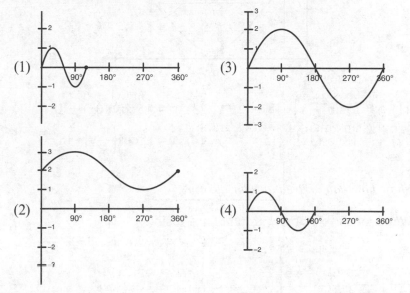

(1)   (3)

(2)   (4)

7. What is the period, in degrees, of the graph of $y = 4\sin\left(\dfrac{2}{3}x\right) + 1$?

(1) 540°   (2) 360°   (3) 270°   (4) 240°

8. What is the period, in radians, of the graph of $y = 4\cos\left(\dfrac{\pi}{10}x\right) - 3$?

(1) 5   (2) 10   (3) 15   (4) 20

9. What could be the equation for the graph if the angles were measured in degrees?

(120°,2)

(1) $y = -2\sin(3x) + 2$ (3) $y = -3\sin(2x) + 2$
(2) $y = 2\sin(3x) + 2$ (4) $y = 3\sin(2x) + 2$

10. What is an equation for this graph if the angle is measured in radians?

(1) $y = 15\sin\left(\dfrac{\pi}{5}x\right) + 18$

(2) $y = 18\cos(3x) + 15$

(3) $y = 15\cos(10x) + 18$

(4) $y = 15\cos\left(\dfrac{\pi}{5}x\right) + 18$

B. *Show how you arrived at your answers.*

1. Sketch one cycle of the graph of $y = 2\sin(x) - 1$.

2. Sketch one cycle of the graph of $y = -3\cos(2x) + 1$.

3. Create an equation for this graph with the angles in radians.

4. Create an equation for this graph with the angles in radians.

5. Sketch a graph of $y = -10\cos\left(\dfrac{\pi}{20}x\right) + 10$.

5.5 GRAPHICALLY SOLVING TRIG EQUATIONS

⋀ KEY IDEAS ⋀

Equations can be solved graphically by using the intersect feature of the graphing calculator. As long as the question doesn't specify that "only an algebraic solution will be accepted," using the graphical method is very quick and accurate.

Section 5.2 described a lengthy process for algebraically solving trig equations. These same problems can be solved very quickly with a graphing calculator.

To find the two solutions between 0° and 360° to $\sin x = -0.94$ using a graphing calculator, use the intersect feature. Graph each side of the equation on the same set of axes, and find the x-coordinate of the intersection points of the two curves.

For this example, graph $y = \sin(x)$ and $y = -0.94$ on the same axes. Make sure the calculator is set to degree mode and that the window ranges from a minimum x-value of 0 or less to a maximum x-value of 360 or more.

For the TI-84:

Press [2nd], [TRACE], [5], [ENTER], [ENTER] to select the sine curve and the horizontal line. Then move near one of the intersection points between 0 and 360. Press [ENTER] again. Do this again for the other intersection point.

For the TI-Nspire:

Press [menu], [6], [4]. Then move to a spot to the left of one of the intersection points between 0 and 360. Press [click]. Then move to a spot to the right of one of the intersection points. Press [click] again.

The process is the same if the angle is in radians, as is the case for the real-world scenarios in the next section. Just remember to change the x-values of the window so you can see from 0 to 2π.

Example

Find the two answers between 0 and 100 to the equation $18 = -10\cos\left(\dfrac{\pi}{50}x\right) + 13$ where the angle is in radians.

Solution: Set the calculator to radian mode. Then graph $y = 18$ and $y = -10\cos\left(\dfrac{\pi}{50}x\right) + 13$. The x-coordinates of the intersection points are approximately $x = 33$ and $x = 67$.

Check Your Understanding of Section 5.5

A. Multiple-Choice

1. Graphically solve the equation $\sin(x) = 0.47$ for $0° \le x < 360°$.
   (1) 28°
   (2) 28° and 332°
   (3) 28° and 208°
   (4) 28° and 152°

2. Graphically solve the equation $\cos(x) = -0.60$ for $0° \le x < 360°$.
   (1) 127° and 233°
   (2) 53° and 127°
   (3) 53° and 307°
   (4) 127° and 307°

3. Graphically solve the equation $\sin(x) = 0.39$ for $0 \le x < 2\pi$.
   (1) 0.4 and 2.7
   (2) 0.5 and 2.8
   (3) 0.6 and 2.9
   (4) 0.7 and 3.0

4. Graphically solve the equation $2\cos(3x) + 3 = 4$ for $0 \le x < \dfrac{2\pi}{3}$.
   (1) 0.25 and 1.65
   (2) 0.30 and 1.70
   (3) 0.35 and 1.75
   (4) 0.40 and 1.80

5. Graphically solve the equation $15\cos\left(\dfrac{\pi}{5}x\right) + 17 = 30$, in radians, for $0 \le x < 10$.
   (1) 0.6 and 9.4
   (2) 0.8 and 9.2
   (3) 1.0 and 9.0
   (4) 1.2 and 8.8

6. Graphically solve the equation $-30\cos\left(\dfrac{\pi}{6}x\right) + 70 = 80$, in radians, for $0 \le x < 12$.
   (1) 3.4 and 8.6
   (2) 3.6 and 8.4
   (3) 3.8 and 8.2
   (4) 4.0 and 8.0

7. Graphically solve the equation $-40\sin\left(\dfrac{\pi}{15}x\right)+35=60$, in radians, for $0 \le x < 30$.
   - (1) 18.2 and 26.8
   - (2) 18.0 and 26.6
   - (3) 17.8 and 26.4
   - (4) 17.6 and 26.2

8. Graphically solve the equation $-30\cos\left(\dfrac{\pi}{25}x\right)+25=10$, in radians, for $0 \le x < 50$.
   - (1) 8.1 and 41.9
   - (2) 8.2 and 41.8
   - (3) 8.3 and 41.7
   - (4) 8.4 and 41.6

9. Graphically solve the equation $15\cos\left(\dfrac{\pi}{20}x\right)+10=3$, in radians, for $0 \le x < 40$.
   - (1) 13 and 27
   - (2) 12 and 28
   - (3) 11 and 29
   - (4) 10 and 30

10. How many solutions are there, for $0° \le x < 360°$, to the equation $\sin(2x) + 1 = \cos(x)$?
    - (1) 0
    - (2) 1
    - (3) 2
    - (4) 3

5.6 MODELING REAL-WORLD SCENARIOS WITH TRIG FUNCTIONS

KEY IDEAS

Many real-world scenarios can be modeled with trig functions. These include models based on spinning circles, bouncing springs, and weather patterns. Most things that go up, down, up, down in a repeating pattern can be modeled with a trig function involving sine or cosine. After creating a function to model a real-world scenario, that function can be used in an equation to calculate things about the model.

The Ferris Wheel Problem

The most common, and easiest to follow, real-world scenario that can be modeled with a sine or cosine curve is the height above the ground of a car on a Ferris wheel.

The radius of this Ferris wheel is 10 feet, and the bottom of the Ferris wheel is 3 feet above the ground. The amount of time the Ferris wheel takes to make a complete revolution is 100 seconds.

When Evelyn gets on the Ferris wheel, she is exactly 3 feet above the ground. Then 50 seconds after starting, she will be 23 feet above the ground, at the peak of the Ferris wheel. Then 100 seconds after starting, she will be back to her starting position, again exactly 3 feet above the ground.

The graph of Evelyn's height above the ground will look like this.

Using the methods from earlier in this chapter, the equation of this curve can be found to be $f(x) = -10\cos\left(\dfrac{\pi}{50}x\right) + 13$.

With this equation, two different kinds of questions can be answered.

1) How high above the ground will Evelyn be after a certain number of seconds?

2) After how many seconds will Evelyn be a certain number of feet above the ground?

To find how high above the ground Evelyn will be after 43 seconds, substitute 43 for x in the equation. $f(43) = -10\cos\left(\dfrac{\pi}{50} \cdot 43\right) + 13 \approx 22$ feet.

To find the two times between 0 and 100 when Evelyn will be 18 feet above the ground, solve the equation $18 = -10\cos\left(\dfrac{\pi}{50}x\right) + 13$. This can be done with algebra or with the graphing calculator. The graphing calculator approach is easiest.

By finding the x-coordinates of the intersection of the graphs of $y = 18$ and $y = -10\cos\left(\dfrac{\pi}{50}x\right) + 13$, you get $x \approx 33$ seconds and $x \approx 67$ seconds.

The Temperature Problem

Often the function for the scenario is already given and all that's left is solving equations with it. In this scenario, a table is provided that contains the temperature in New York over the course of two years.

Year	Temperature in NY (°F)										
2013	35.1	33.9	40.1	53.0	62.8	72.7	79.8	74.6	67.9	60.2	45.3	38.5
2014	28.6	31.6	37.7	52.3	64.0	72.5	76.1	74.5	69.7	59.6	45.3	40.5

A function that approximately models this data is $f(x) = -23\cos\left(\dfrac{\pi}{6}x\right) + 54$.

Use this function to answer the two types of questions.

Question 1: If the beginning of January is month $x = 0$, what is the temperature in the middle of March ($x = 2.5$)?

Substitute $x = 2.5$ into the function. $f(2.5) = -23\cos\left(\dfrac{\pi}{6} \cdot 2.5\right) + 54 \approx 48$ degrees

Question 2: At what two times is the temperature approximately 70 degrees?

Graph $y = 70$ and $y = -23\cos\left(\dfrac{\pi}{6}x\right) + 54$ on the graphing calculator. Find the x-coordinates of the two intersection points.

They intersect at $x = 4.5$ and $x = 7.5$. These x-values correspond to the middle of May and the middle of August.

Example

In New York on March 20 (which is the spring equinox), there are 12 hours of daylight and 12 hours of darkness. Approximately 91 days later on June 20 (which is the summer solstice), there are 15 hours of daylight and 9 hours of darkness. Approximately 91 days after that on September 22 (which is the fall equinox), there are 12 hours of daylight and 12 hours of darkness again. Approximately 91 days after that on December 21 (which is the winter solstice), there are 9 hours of daylight and 15 hours of darkness. Approximately 91 days after that, it is the spring equinox again.

Sketch the graph of a function $f(x)$ where x is the number of days that has passed since the spring equinox and $f(x)$ is the number of hours of daylight on that day. Find an approximate equation for this function.

Solution: The graph looks like this.

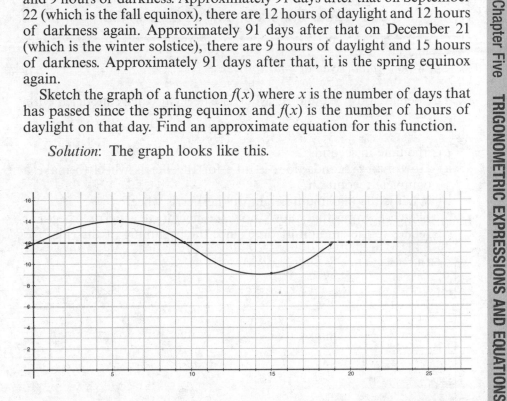

The equation is a sine curve with a positive A-value. The C-value is $+12$ since $y = 12$ is the axis of the sine curve. The A-value is $+3$ since the high point is 3 units above the axis and the low point is 3 units below the axis.

The B-value can be solved with the equation $364 = \dfrac{2\pi}{B}$: $364B = 2\pi$, $B = \dfrac{2\pi}{364}$, $B \approx \dfrac{\pi}{182}$.

The function is approximately $f(x) = 3\sin\left(\dfrac{\pi}{182}x\right) + 12$.

Check Your Understanding of Section 5.6

B. *Show how you arrived at your answers.*

1. The height above ground of a Ferris wheel car can be modeled with the equation $h = -20\cos\left(\frac{\pi}{15}t\right) + 24$, where h is the height in feet and t is the time in seconds.
   a) How many seconds does it take for the Ferris wheel to make a complete revolution?
   b) What is the maximum height of the Ferris wheel?
   c) Sketch a graph that shows how the height of a Ferris wheel car relates to the amount of time since the beginning of the ride.

2. Matthew gets on a Ferris wheel at the bottom of the wheel, which is 5 feet above the ground. After 40 seconds, he is at the top of the wheel, 65 feet high.
   a) Sketch a graph that shows the relationship between time and height.

b) Create an equation that relates h and t.
c) Determine how high up Matthew will be after 55 seconds.
d) Determine the two times between 0 and 60 seconds when Matthew is 41 feet above the ground.

3. Below is the graph of the average temperature in Regentsville in each of the months where 0 = January, 1 = February, 2 = March, 3 = April, 4 = May, 5 = June, 6 = July, 7 = August, 8 = September, 9 = October, 10 = November, 11 = December, and 12 = next January.

a) Create an equation that relates temperature (y) and month (x).
b) Use that equation to find out the temperature in March ($x = 2$).
c) Use that equation to find the two times when the temperature is 70 degrees.

5.7 TRIGONOMETRY IDENTITIES

KEY IDEAS

A *trig identity* is an equation in which the left side and the right side both contain trig expressions. Complicated trig expressions can sometimes be simplified by replacing parts of it with equivalent expressions and then simplifying the new expression.

MATH FACTS

The expression $(\sin x)^2$ is more commonly written as $\sin^2 x$. This is to prevent it from being confused with $\sin(x^2)$. The same applies to $(\cos x)^2$. It is more commonly written as $\cos^2 x$.

There are five main trig identities.

1. $\sin^2 x + \cos^2 x = 1$ or

   $\sin^2 x = 1 - \cos^2 x$ or

   $\cos^2 x = 1 - \sin^2 x$

2. $\tan x = \dfrac{\sin x}{\cos x}$

3. $\csc x = \dfrac{1}{\sin x}$

4. $\sec x = \dfrac{1}{\cos x}$

5. $\cot x = \dfrac{1}{\tan x}$

Trig identities can be used to identify trig expressions that are equivalent to other trig expressions.

Rewrite All Expressions in Terms of Sine and Cosine

One way to simplify a trig expression is first to use the five basic identities described earlier to rewrite each expression in terms of sine and cosine. After this is done, you may be able to simplify further by reducing the fraction that remains.

The expression $\dfrac{\tan x}{\sec x}$ can be shown to be equivalent to $\sin x$ with the following steps:

$$\frac{\tan x}{\sec x}$$

$$\frac{\dfrac{\sin x}{\cos x}}{\dfrac{1}{\cos x}}$$

$$\frac{\sin x}{\cancel{\cos x}} \cdot \frac{\cancel{\cos x}}{1}$$

$$\sin x$$

Example 1

Show the steps to demonstrate that $\dfrac{\sec x}{\csc x}$ is equivalent to $\tan x$.

Solution:

$$\frac{\sec x}{\csc x}$$

$$\frac{\dfrac{1}{\cos x}}{\dfrac{1}{\sin x}}$$

$$\frac{1}{\cos x} \cdot \frac{\sin x}{1}$$

$$\frac{\sin x}{\cos x}$$

$$\tan x$$

Example 2

Show that $\tan x \cdot \csc x = \sec x$.

Solution:

$$\tan x \cdot \csc x$$

$$\frac{\sin x}{\cos x} \cdot \frac{1}{\sin x}$$

$$\frac{\cancel{\sin x}}{\cos x \cdot \cancel{\sin x}}$$

$$\frac{1}{\cos x}$$

$$\sec x$$

Combining Fractions in Trig Identity Problems

Sometimes in a trig identity problem, fractions must be combined. In this case, a common denominator needs to be found. Often when the numerators are later combined, another trig identity helps further simplify the expression.

Example 3

Show that $\sin x \cdot \tan x + \cos x = \sec x$.

Solution:

$$\sin x \cdot \tan x + \cos x$$

$$\sin x \cdot \frac{\sin x}{\cos x} + \cos x$$

$$\frac{\sin^2 x}{\cos x} + \frac{\cos x}{1}$$

Here the common denominator for the two fractions is $\cos x$.

$$\frac{\sin^2 x}{\cos x} + \frac{\cos x \cdot \cos x}{\cos x}$$

$$\frac{\sin^2 x + \cos^2 x}{\cos x}$$

The first trig identity can be used to simplify the numerator.

$$\frac{1}{\cos x}$$

$$\sec x$$

More Complicated Trig Identities

A more complicated type of trig identity relies on the fact that an identity is the difference of perfect squares. For instance, $\cos^2 x = 1 - \sin^2 x$ can be factored into $(1 - \sin x)(1 + \sin x)$. Similarly, $\sin^2 x = 1 - \cos^2 x = (1 - \cos x)(1 + \cos x)$.

Example 4

Show that $\dfrac{1+\cos x}{\sin x}$ is equivalent to $\dfrac{\sin x}{1-\cos x}$.

Solution: Notice that the $1 + \cos x$ expression is one of the factors of $(1 - \cos x)(1 + \cos x) = 1 - \cos^2 x = \sin^2 x$.

As a first step, you should reduce the original fraction by multiplying both the numerator and denominator by $1 - \cos x$.

$$\frac{1 + \cos x}{\sin x}$$

$$\frac{(1 + \cos x)(1 - \cos x)}{\sin x(1 - \cos x)}$$

$$\frac{1 - \cos^2 x}{\sin x(1 - \cos x)}$$

$$\frac{\sin^2 x}{\sin x(1 - \cos x)}$$

$$\frac{\cancel{\sin x} \cdot \sin x}{\cancel{\sin x}(1 - \cos x)}$$

$$\frac{\sin x}{1 - \cos x}$$

Check Your Understanding of Section 5.7

A. *Multiple-Choice*

1. $\csc\theta$ is equal to which of the following?

(1) $\dfrac{1}{\cos\theta}$ (3) $\dfrac{1}{\sin\theta}$

(2) $\dfrac{1}{\tan\theta}$ (4) $\dfrac{1}{\sec\theta}$

2. Which is *not* a true trig identity?
  (1) $\sin^2\theta + \cos^2\theta = 1$ (3) $1 - \sin^2\theta = \cos^2\theta$
  (2) $\sin^2\theta - \cos^2\theta = 1$ (4) $1 - \cos^2\theta = \sin^2\theta$

3. $\cos\theta \cdot \sec\theta$ is equal to which of the following?
(1) $\tan\theta$ (3) $\sec\theta$
(2) $\cot\theta$ (4) 1

4. $\tan\theta(1 - \sin^2\theta)$ is equal to which of the following?
(1) $\sin\theta\cos\theta$ (3) $\csc\theta\tan\theta$
(2) $\sin\theta\sec\theta$ (4) $\sec\theta\tan\theta$

5. $\dfrac{1}{\cos\theta} - \cos\theta$ is equal to which of the following?
(1) $\tan\theta\sin\theta$ (3) $\cos\theta\cot\theta$
(2) $\cot\theta\sin\theta$ (4) $\sec\theta\sin\theta$

6. $\cos^2\theta - \sin^2\theta$ is equal to which of the following?
(1) $2\cos^2\theta - 1$ (3) -1
(2) $2\cos^2\theta + 1$ (4) 1

7. $\dfrac{\cot\theta}{1 + \sin\theta}$ is equal to which of the following?

(1) $\dfrac{1 + \sin\theta}{\cos\theta\sin\theta}$ (3) $\dfrac{1 + \cos\theta}{\cos\theta\sin\theta}$

(2) $\dfrac{1 - \cos\theta}{\cos\theta\sin\theta}$ (4) $\dfrac{1 - \sin\theta}{\cos\theta\sin\theta}$

8. $\cos\theta$ is equal to which of the following?
(1) $\sin(180° - \theta)$ (3) $\sin(\theta - 90°)$
(2) $\sin(\theta + 90°)$ (4) $\sin(90° - \theta)$

9. $-\tan\theta\sin\theta$ is equal to which of the following?
(1) $\sec\theta - \cos\theta$ (3) $\sin\theta - \csc\theta$
(2) $\cos\theta - \sec\theta$ (4) $\csc\theta - \sin\theta$

10. If $0° \le \theta < 90°$ and $\csc\theta = \dfrac{5}{3}$, what is the value of $\cot\theta$?

(1) $\dfrac{5}{4}$ (2) $\dfrac{4}{3}$ (3) $\dfrac{4}{5}$ (4) $\dfrac{3}{4}$

B. *Show how you arrived at your answers.*

1. Prove that $\sin \theta \tan \theta + \cos \theta = \sec \theta$.

2. Prove that $\dfrac{\cos \theta}{1 + \sin \theta} = \dfrac{1 - \sin \theta}{\cos \theta}$.

3. Prove that $(\cos \theta - \sin \theta)(\cos \theta + \sin \theta) + 2\sin^2 \theta = 1$.

4. Prove that $(\cot \theta + \csc \theta)(\cot \theta - \csc \theta) = -1$.

5. What identity do you get if you divide both sides of the identity $\sin^2 \theta + \cos^2 \theta = 1$ by $\sin^2 \theta$?

SYSTEMS OF EQUATIONS

6.1 TWO EQUATIONS WITH TWO UNKNOWNS

⎯⎯⎯⎯⎯⎯⎯⎯⎯⎯⎯ △ ⎯⎯⎯⎯⎯⎯⎯⎯⎯⎯⎯
KEY IDEAS
△⎯⎯△

A *system of two equations with two unknowns* is two equations each
containing two variables. The solution set to a system of two equations
with two unknowns is usually an *ordered pair* that makes both equations
true.

The ordered pair (3, 7) is a solution to the equation $x + y = 10$. If the
first number is substituted for the x and if the second number is substituted for the y, the equation becomes $3 + 7 = 10$, which is true. Other
ordered pairs that are solutions to this equation are (8, 2), (5, 5), (10, 0),
and (12, –2).

An example of a system of two equations with two unknowns is

$$x + y = 10$$
$$x - y = 4$$

The goal is to find, if one exists, an ordered pair that satisfies both equations at the same time. By examining the different ordered pairs that
make the first equation true and by testing to see if they also make the
second equation true, it is possible through the process of guess-and-
check to find that the ordered pair (7, 3) is the solution because $7 + 3 =
10$ and $7 - 3 = 4$. For more complicated examples, though, a method that
uses algebra is needed.

The Substitution Method

If one of the variables is already isolated in one of the equations, it is
possible to simplify the two equations with two variables into just one
equation with one variable.

$$y = 3x - 2$$
$$x + 2y = 24$$

Since the y is isolated in the first equation, the y in the second equation can be replaced with $(3x - 2)$.

$$x + 2(3x - 2) = 24$$
$$x + 6x - 4 = 24$$
$$7x - 4 = 24$$
$$\underline{+4 = +4}$$
$$7x = 28$$
$$\frac{7x}{7} = \frac{28}{7}$$
$$x = 4$$

To get the other number in the ordered pair, substitute the $x = 4$ into the first equation.

$$y = 3 \cdot 4 - 2 = 12 - 2 = 10$$

The ordered pair that satisfies both equations is (4, 10).

Example 1

Find the solution set of the system of equations.

$$y = 2x - 5$$
$$5x - 3y = 9$$

Solution: Substitute $(2x - 5)$ for y in the second equation.

$$5x - 3(2x - 5) = 9$$
$$5x - 6x + 15 = 9$$
$$-x + 15 = 9$$
$$\underline{-15 = -15}$$
$$-x = -6$$
$$\frac{-x}{-1} = \frac{-6}{-1}$$
$$x = 6$$

Substitute $x = 6$ into one of the original equations.

$$y = 2 \cdot 6 - 5 = 12 - 5 = 7$$

The solution is (6, 7).

The Elimination Method

If the two equations can be combined in a way that *eliminates* one of the variables, the remaining equation can be solved for one part of the ordered pair. Sometimes this method requires that you multiply both sides of one or of both equations by some constant.

Case 1:

A coefficient of one variable is the opposite of the coefficient of the same variable in the other equation.

Examine the coefficients of the system of equations.

$$2x + 3y = 11$$
$$x - 3y = 1$$

The coefficient of the y in the first equation is 3. The coefficient of the y in the second equation is -3. Remember that $3 + -3 = 0$. If these equations are added together, the result will be an equation with no y-term.

$$2x + 3y = 11$$
$$+ \quad x - 3y = 1$$
$$\overline{3x = 12}$$
$$\frac{3x}{3} = \frac{12}{3}$$
$$x = 4$$

Substitute $x = 4$ into either original equation and solve for y.

$$4 - 3y = 1$$
$$\underline{-4 = -4}$$
$$-3y = -3$$
$$\frac{-3y}{-3} = \frac{-3}{-3}$$
$$y = 1$$

The solution to the system of equations is the ordered pair (4, 1).

Case 2:

A coefficient of one of the variables in one of the equations is equal to the coefficient of the same variable in the other equation.

Examine the coefficients in the system of equations.

$$4x + 3y = 29$$

$$4x - 2y = 14$$

The coefficients of the x-terms in both equations are equal to $+4$.

It is permitted to multiply both sides of an equation by the same constant. For this situation, multiply both sides of either equation by -1. Below is what happens if both sides of the second equation are multiplied by -1.

$$4x + 3y = 29$$

$$-1(4x - 2y) = -1(14)$$

$$4x + 3y = 29$$

$$-4x + 2y = -14$$

Now this has become like Case 1.

$$\begin{array}{r} 4x + 3y = 29 \\ +\ -4x + 2y = -14 \\ \hline 5y = 15 \end{array}$$

$$\frac{5y}{5} = \frac{15}{5}$$

$$y = 3$$

Substitute 3 for y in either of the original equations.

$$4x + 3 \cdot 3 = 29$$

$$4x + 9 = 29$$

$$4x = 20$$

$$x = 5$$

The solution set is (5, 3).

Case 3:

A coefficient of one of the variables in one of the equations is a multiple of the same variable in the other equation.

Examine the coefficients of the system of equations.

$$2x + 6y = 4$$

$$5x + 3y = 34$$

231

The coefficient of the y in the first equation, 6, is a multiple of the coefficient of the y in the second equation, 3. For the equation that has the smaller coefficient, multiply both sides of the equation by the number that would make that coefficient the opposite of the one from the other equation. For this example, multiply both sides of the second equation by -2.

$$2x + 6y = 4$$

$$-2(5x + 3y) = -2(34)$$

$$2x + 6y = 4$$

$$+ \quad -10x - 6y = -68$$

$$-8x = -64$$

$$\frac{-8x}{-8} = \frac{-64}{-8}$$

$$x = 8$$

Now substitute $x = 8$ into one of the original equations.

$$2 \cdot 8 + 6y = 4$$

$$16 + 6y = 4$$

$$-16 = -16$$

$$6y = -12$$

$$\frac{6y}{6} = \frac{-12}{6}$$

$$y = -2$$

The solution is $(8, -2)$.

Case 4:

No coefficient for either variable in either equation is a multiple of the coefficient of the same variable in the other equation.

Examine the coefficients of the system of equations.

$$3x + 4y = 42$$

$$5x - 6y = 32$$

For the x-terms, the 5 is not a multiple of 3. For the y-terms, the -6 is not a multiple of the 4. When this happens, both equations must be changed.

You can choose whether you want to change the equations so that either the x-variables will be eliminated or the y-variables will be eliminated.

To eliminate the y-variables, find the least common multiple of the two coefficients. In this case, the least common multiple of 4 and 6 is 12. The goal is to convert one of the y-coefficients into a +12 and the other into a −12. This can be done by multiplying both sides of the first equation by +3 and both sides of the second equation by +2.

$$3(3x + 4y) = 3(42)$$

$$2(5x - 6y) = 2(32)$$

$$9x + 12y = 126$$

$$+ \quad 10x - 12y = 64$$

$$\overline{ 19x = 190}$$

$$\frac{19x}{19} = \frac{190}{19}$$

$$x = 10$$

Substitute 10 for x into either of the original equations.

$$3 \cdot 10 + 4y = 42$$

$$30 + 4y = 42$$

$$\underline{-30 = -30}$$

$$4y = 12$$

$$\frac{4y}{4} = \frac{12}{4}$$

$$y = 3$$

The solution is (10, 3).

Example 2

Solve the system of equations by eliminating the x-terms.

$$3x + 7y = 4$$

$$4x - 5y = 34$$

Solution: The least common multiple of the x-coefficients is 12. Multiply both sides of the first equation by +4 and both sides of the second equation by −3.

$$4(3x + 7y) = 4(4)$$
$$-3(4x - 5y) = -3(34)$$
$$12x + 28y = 16$$
$$+\quad -12x + 15y = -102$$
$$\overline{ 43y = -86}$$
$$\frac{43y}{43} = \frac{-86}{43}$$
$$y = -2$$

Substitue $y = -2$ into one of the original equations.

$$3x + 7(-2) = 4$$
$$3x - 14 = 4$$
$$\underline{+14 = +14}$$
$$3x = 18$$
$$\frac{3x}{3} = \frac{18}{3}$$
$$x = 6$$

The solution is $(6, -2)$.

MATH FACTS

Sometimes when solving a system of equations with the elimination, both variables get eliminated and the resulting equation has just numbers. If the two numbers of the resulting equation are not equal, there is no solution to the original system. If the two numbers of the resulting equation are equal, there are an infinite number of solutions to the equation.

Example 3

How many solutions does this system of equations have?

$$2x + 3y = -10$$
$$-2x - 3y = 12$$

(1) No solutions
(2) One solution
(3) Two solutions
(4) More than two solutions

Solution: When the two equations are added together, the result is $0 = 2$, which is never true. There are no solutions to the original system of equations. Choice (1) is correct.

Linear-Quadratic Systems of Equations

If one of the equations has an exponent of 2, the system can be solved with the substitution method. The resulting equation will be a quadratic equation with up to two solutions.

Example 4

What ordered pair(s) satisfy the system?

$$x = y - 1$$
$$y = x^2 - 1$$

Solution:

$$y = (y - 1)^2 - 1$$
$$y = y^2 - 2y + 1 - 1$$
$$y = y^2 - 2y$$
$$0 = y^2 - 3y$$
$$0 = y(y - 3)$$
$$y = 0 \text{ or } y = 3$$

For $y = 0$, $x = -1$. For $y = 3$, $x = 2$. So the two solutions are $(-1, 0)$ and $(2, 3)$.

Check Your Understanding of Section 6.1

A. Multiple-Choice

1. Solve the system of equations.

$$y = 2x + 1$$
$$3x + 4y = 15$$

   (1) (3, 1)   (2) (4, 2)   (3) (2, 4)   (4) (1, 3)

2. Solve the system of equations.

$$y = 3x - 2$$
$$7x - 2y = 7$$

   (1) (3, 7)   (2) (7, 3)   (3) (4, 6)   (4) (6, 4)

3. Solve the system of equations.

$$3x + 5y = 37$$
$$2x - 5y = 8$$

   (1) (9, 2)   (2) (2, 9)   (3) (8, 3)   (4) (3, 8)

4. Solve the system of equations.

$$4x + 3y = -11$$
$$-4x + 5y = 35$$

   (1) (5, -3)   (2) (-3, 5)   (3) (-5, 3)   (4) (3, -5)

5. Solve the system of equations.

$$2x + 3y = 11$$
$$3x - 6y = 6$$

   (1) (4, 1)   (2) (1, 4)   (3) (2, 3)   (4) (3, 2)

6. Solve the system of equations.

$$3x - 4y = 7$$
$$5x + 3y = 31$$

   (1) (2, 5)   (2) (5, 2)   (3) (3, 4)   (4) (4, 3)

7. Solve the system of equations.

$$5x - 3y = 27$$
$$6x - 7y = 46$$

(1) (3, −4)   (2) (−4, 3)   (3) (−3, 4)   (4) (4, −3)

8. Solve the system of equations.

$$2x - 5y = 7$$
$$4x - 10y = 12$$

(1) (6, 1)   (2) (11, 3)   (3) (1, 6)   (4) No solution

9. How many solutions does the system of equations have?

$$5x + 3y = 10$$
$$10x + 6y = 20$$

(1) No solution   (3) Two solutions
(2) One solution   (4) Infinite solutions

10. If a system of two equations with two unknowns has no solution, what do the graphs of the two lines representing the two equations look like?
(1) They intersect at one point.
(2) They are parallel.
(3) They are the same line.
(4) They intersect at two points.

B. *Show how you arrived at your answers.*

1. When will it be easier to use the substitution method rather than the elimination method when solving a system of two equations with two unknowns?

2. Reginald is solving the system of equations.

$$x + y = 12$$
$$x - y = 8$$

Reginald adds the two equations to get $2x = 20$. Madelyn subtracts the two equations to get $2y = 4$. Who is correct?

3. Solve the system of equations.

$$3x + 5y = -2$$
$$4x + 3y = -10$$

4. Movie tickets cost $7 for children and $12 for adults. If 10 people purchase tickets and the total price is $85, how many adult tickets and how many children tickets were purchased?

5. The following system of equations has the solution (4, 7).

$$2x + y = m$$
$$3x - y = n$$

What are the values of m and n?

6.2 THE ELIMINATION METHOD FOR THREE EQUATIONS WITH THREE UNKNOWNS

KEY IDEAS

With three equations and three unknowns, the solution can be an *ordered triple* (x, y, z). The process of solving a system of equations with three unknowns requires three different elimination steps.

A system of three equations with three unknowns looks like this.

$$-3x - 7y - 2z = -31$$
$$5x + 4y + 2z = 35$$
$$-x + y - z = -4$$

Step 1:

Decide which variable to eliminate first.

The first step in solving a system of three equations with three unknowns is to examine the coefficients of each of the variables to see if the variables for any coefficient are matching numbers or have numbers that are multiples of the other coefficients. Any variable can be eliminated. However, it is easier to eliminate a variable when the coefficients contain multiples of other coefficients.

In this example, the z-variables have the coefficients -2, $+2$, and -1. The z will be the variable we try to eliminate.

Step 2:

Combine the first and second equation to eliminate a variable.

Since -2 is the opposite of $+2$, this is like Case 1 from the previous section. The z can be eliminated by adding the two equations together. This step could require multiplying both sides of one or both equations by a constant before combining them.

$$
\begin{array}{r}
-3x - 7y - 2z = -31 \\
+ \quad 5x + 4y + 2z = 35 \\
\hline
2x - 3y = 4
\end{array}
$$

This equation $2x - 3y = 4$ will be used in Step 4.

Step 3:

Combine the second and third equation (or the first and third if it is easier) to eliminate a variable.

Since $+2z + -z$ will not eliminate z, both sides of the second equation must first be multiplied by $+2$ so the z-coefficient will become -2, the opposite of the coefficient of the other z, $+2$.

$$5x + 4y + 2z = 35$$
$$2(-x + y - z) = 2(-4)$$

Now the equations can be combined to eliminate z.

$$
\begin{array}{r}
5x + 4y + 2z = 35 \\
+ \quad -2x + 2y - 2z = -8 \\
\hline
3x + 6y = 27
\end{array}
$$

This equation, together with the one from Step 2, is used in Step 4.

Step 4:

Solve the resulting system of two equations with two unknowns

$$2x - 3y = 4$$
$$3x + 6y = 27$$

Since the +6 is a multiple of the −3, multiply both sides of the first equation by +2.

$$2(2x - 3y) = 2(4)$$
$$3x + 6y = 27$$

$$4x - 6y = 8$$
$$+ \quad 3x + 6y = 27$$
$$\overline{7x = 35}$$
$$\frac{7x}{7} = \frac{35}{7}$$
$$x = 5$$

Substitute 5 for x into either of the two-variable equations.

$$2(5) - 3y = 4$$
$$10 - 3y = 4$$
$$\underline{-10 = -10}$$
$$-3y = -6$$
$$\frac{-3y}{-3} = \frac{-6}{-3}$$
$$y = 2$$

Step 5:

Substitute the solution of the two-variable system of equations into one of the original equations to find the solution for the third variable.

Any of the original three equations can be used. Since the third equation has the smallest coefficients, it will be simplest to use.

$$-(5) + 2 - z = -4$$
$$-5 + 2 - z = -4$$
$$-3 - z = -4$$
$$\underline{+3 = +3}$$
$$-z = -1$$
$$\frac{-z}{-1} = \frac{-1}{-1}$$
$$z = 1$$

The solution is the ordered triple (5, 2, 1).

This can be checked by substituting $x = 5$, $y = 2$, $z = 1$ into each of the three original equations.

$$-3 \cdot 5 - 7 \cdot 2 - 2 \cdot 1 \overset{?}{=} -31$$
$$-15 - 14 - 2 \overset{?}{=} -31$$
$$-31 \overset{\checkmark}{=} -31$$
$$5 \cdot 5 + 4 \cdot 2 + 2 \cdot 1 \overset{?}{=} 35$$
$$25 + 8 + 2 \overset{?}{=} 35$$
$$35 \overset{\checkmark}{=} 35$$
$$-5 + 2 - 1 \overset{?}{=} -4$$
$$-4 \overset{\checkmark}{=} -4$$

Example

Find the solution to the system of equations.

$$-x + 5y + z = -18$$
$$4x - 5y + 3z = 34$$
$$x + 5y - z = -12$$

Solution: Since the *y*-variables have coefficients of +5, −5, and +5, the *y* should be eliminated. Combining equations 1 and 2 gets $3x + 4z = 16$. Combining equations 2 and 3 gets $5x + 2z = 22$.

Now solve the system.

$$3x + 4z = 16$$
$$5x + 2z = 22$$

Multiply both sides of the second equation by −2 so the *z*-coefficient becomes −4.

$$3x + 4z = 16$$
$$+ \quad -10x - 4z = -44$$
$$\overline{\quad -7x = -28 \quad}$$
$$x = 4$$

Substitute $x = 4$ into one of the two-variable equations.

$$3 \cdot 4 + 4z = 16$$
$$12 + 4z = 16$$
$$4z = 4$$
$$z = 1$$

Now substitute $x = 4$ and $z = 1$ into one of the original equations.

$$-4 + 5y + 1 = -18$$
$$-3 + 5y = -18$$
$$\underline{+3 = +3}$$
$$5y = -15$$
$$y = -3$$

The solution is (4, −3, 1).

Check Your Understanding of Section 6.2

B. Show how you arrived at your answers.

1. If $x = 2$ and $y = 1$ and if $3x + 2y + 4z = 24$, what is the value of z?

2. Solve the system of equations.

$$3x + 2y + z = 15$$
$$2y + z = 6$$
$$z = 4$$

3. Solve the system of equations.

$$5x + 3y + z = 23$$
$$3x + 4y - z = 21$$
$$4x + 5y - 2z = 26$$

4. Solve the system of equations.

$$4x - 2y + 3z = 27$$
$$2x + 3y - 2z = -1$$
$$5x - 7y + 4z = 39$$

5. Solve the system of equations.

$$3x - 4y + 5z = -38$$
$$4x - 3y + 6z = -41$$
$$5x - 3y - 7z = 9$$

6.3 USING A GRAPHING CALCULATOR TO SOLVE SYSTEMS OF EQUATIONS

The *matrix* function of the graphing calculator can solve systems of equations instantly. If the system of equations is a multiple-choice question, using the calculator method will earn full credit. If the question requires an algebraic solution, using the calculator is still a useful way to check your answer.

Solving Two Equations with Two Unknowns Using Matrices

Matrices is an advanced math topic usually taught in precalculus. Even without understanding what matrices are or how they work, you can use them to solve systems of equations with a graphing calculator.

In Section 6.1, algebra was used to solve the following system of equations.

$$3x + 7y = 4$$

$$4x - 5y = 34$$

To solve the same system with the graphing calculator, do the following.

For the TI-84:

Press [2ND] [x^{-1}] to get to the matrix menu. Move to "EDIT," and press [1] to edit matrix A. Change the numbers before and after the "x" to 2 and 3. Put the numbers 3, 7, and 4 into row 1 and the numbers 4, −5, and 34 into row 2. Press [2ND] [MODE] to quit the matrix menu.

```
NAMES MATH EDIT
1▉[A]
2: [B]
3: [C]
4: [D]
5: [E]
6: [F]
7↓[G]
```

```
MATRIX[A]  2 ×3
[ 3    7    4 ]
[ 4   -5   34 ]


2,3=34
```

Press [2ND] [x⁻¹] and move to "MATH." Scroll to choice "B," called "rref" (Don't ask!) and select it. Then press [2ND] [x⁻¹] [1] [ )] and press [ENTER].

The numbers in the third column are the values for x and y.

The solution is $(6, -2)$.

For the TI-Nspire:

Open the "Calculate" Scratchpad by pressing [A] from the home screen. Press [menu] [7] [5] for the "Reduced Row-Echelon Form" function. Press [template]. Select the matrix icon in the second row of symbols, sixth from the right. Set the number of rows to 2 and the number of columns to 3. Enter 3, 7, and 4 into the first row and 4, −5, and 34 into the second row. Press [enter].

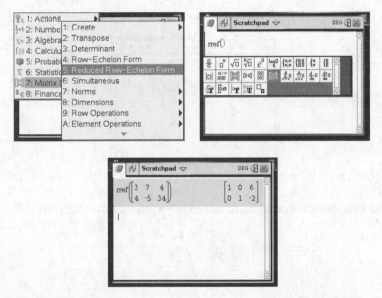

Solving Three Equations with Three Unknowns Using Matrices

Using a graphing calculator to solve a system with three equations and three unknowns is nearly identical to the process for solving two equations with two unknowns. The only difference is that instead of using a matrix with two rows and three columns, use a matrix with three rows and four columns.

This was an example from Section 6.2 that was solved with algebra.

$$-3x - 7y - 2z = -31$$

$$5x + 4y + 2z = 35$$

$$-x + y - z = -4$$

For the TI-84:

Make the matrix with 3 rows and 4 columns.

For the TI-Nspire:

Make the matrix with 3 rows and 4 columns.

Example

Use the graphing calculator to solve the system of equations.

$$2x + 4y + z = 9$$

$$3x + 5y + 6z = -4$$

$$5x + y + 8z = -30$$

Solution: The solution is (−2, 4, −3).

FUNCTIONS

7.1 COMPOSITE FUNCTIONS

KEY IDEAS

A *function* is a mathematical rule that generally takes a number as an input and then outputs a number. If the function is called *f*, the notation $f(4) = 11$ means that when the number 4 is put into the function, the number 11 is output from the function. Functions can be used to describe mathematical relationships, including real-world scenarios.

Functions are often defined by a formula. For example, the formula $f(x) = 2x + 3$ means that when a number is put into the function, a number that is three more than twice that number will come out of the function. So $f(4) = 2 \cdot 4 + 3 = 8 + 3 = 11$.

Not only can numbers be put into a function but so can variables or even other functions. Using the function *f* defined above, $f(a) = 2a + 3$ and $f(x^2 + 1) = 2(x^2 + 1) + 3 = 2x^2 + 2 + 3 = 2x^2 + 5$.

If another function *g* is defined as $g(x) = 3x - 2$, it is possible to create a new function $f(g(x))$. By putting the $g(x)$ into the *f* function, the new function becomes $f(g(x)) = 2g(x) + 3 = 2(3x - 2) + 3$, which can be further simplified to $6x - 4 + 3 = 6x - 1$.

When a function is put into another function, the result is called a *composite function*. When working with composite functions, start with the inner function first and then move to the outer function.

Example 1

If $f(x) = 5x + 2$ and $g(x) = 3x - 1$, what is the value of $f(g(4))$?

Solution:

Since $g(4) = 3 \cdot 4 - 1 = 11$, $f(g(4)) = f(11) = 5 \cdot 11 + 2 = 57$.

Example 2

If $f(x) = 2x + 3$ and $g(x) = 3x - 2$, what are $g(f(x))$ and $f(g(x))$?

Solution:

$$g(f(x)) = g(2x + 3) = 3(2x + 3) - 2 = 6x + 9 - 2 = 6x + 7$$

$$f(g(x)) = f(3x - 2) = 2(3x - 2) + 3 = 6x - 4 + 3 = 6x - 1$$

Notice that in this example, $g(f(x))$ is not equivalent to $f(g(x))$.

Example 3

If $f(x) = \dfrac{x-1}{3}$ and $g(x) = 3x + 1$, what is $f(g(x))$?

Solution:

$$f(g(x)) = f(3x + 1) = \frac{3x + 1 - 1}{3} = \frac{3x}{3} = x$$

More difficult than finding the composition of two given functions is trying to find two functions whose composition would become some given function. For example

$h(x) = 5x - 3$ could be *decomposed* into $h(x) = f(g(x))$ where $g(x) = 5x$ and $f(x) = x - 3$. There are other ways to decompose this function into two functions also, but this way is the most useful.

Example 4

Which of the following could be $f(x)$ and $g(x)$ if $f(g(x)) = (x - 2)^2$?

(1) $f(x) = x - 2$, $g(x) = x^2$
(2) $f(x) = x^2$, $g(x) = x - 2$
(3) $f(x) = x + 2$, $g(x) = x^2$
(4) $f(x) = x^2$, $g(x) = x + 2$

Solution: Choice (2) is correct since $f(x - 2) = (x - 2)^2$. If you test choice (1), it would become $f(x^2) = x^2 - 2$.

Composite Functions on the Graphing Calculator

Graphing calculators can also evaluate functions and composite functions. Using either the TI-84 or the TI-Nspire, this is how to evaluate the function $g(x) = 3x - 1$ at $x = 4$.

For the TI-84:

Press [Y=] and then enter $3x - 1$ after "Y1=." Press [2ND] [MODE] to exit that menu. Press [VARS] [RIGHT] [1], and select [1] for the Y1 function. Press [(] [4] [)] and [ENTER].

For the TI-Nspire:

From the home screen, press [A] for the calculator scratchpad. Press [menu], [1], and [1] for Define. Type $g(x) = 3x - 1$ and press [enter]. To evaluate the function at the value 4, type $g(3)$ and press [enter].

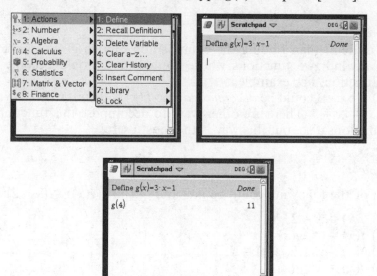

To evaluate a composite function like $f(g(4))$ when $f(x) = 5x + 2$ and $g(x) = 3x - 1$, both functions need to be entered into the calculator. For the TI-84, enter them into Y1 and Y2. For the TI-Nspire, enter them into $f(x)$ and $g(x)$.

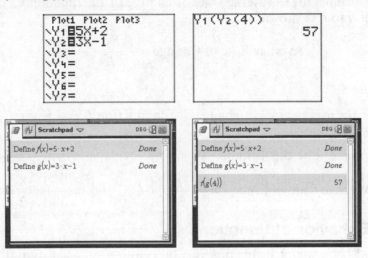

Odd and Even Functions

A function is called an *even* function if its graph has y-axis symmetry. A function is called an *odd* function if its graph has origin symmetry. The graphs of most functions have neither of these symmetries and so are neither odd nor even.

Average Rate of Change

A function has something called *the average rate of change on an interval*. This is very similar to the concept of slope discussed in Chapter 1.

The formula for the average rate of change of the function $f(x)$ on the interval a to b is shown:

$$\text{Average Rate of Change} = \frac{f(b) - f(a)}{b - a}$$

Example 5

What is the average rate of change of the function $f(x) = x^2$ on the interval $x = 1$ to $x = 3$?

Solution: Use the formula:

$$\text{Average Rate of Change} = \frac{f(3) - f(1)}{3 - 1} = \frac{3^2 - 1^2}{3 - 1} = \frac{9 - 1}{3 - 1} = \frac{8}{2} = 4$$

End Behavior of Functions

The end behavior of a function is a description of what output values happen when very large positive or very large negative values are put into the function. When a large number is put into the function $f(x) = x^2$, for example $f(1,000) = 1,000^2 = 1,000,000$, the function outputs a very large positive number.

Using symbols, we write: As $x \to \infty$, $f(x) \to \infty$, where the ∞ means infinity.

When a large negative number is put into the same function, like $f(-1,000) = (-1,000)^2 = +1,000,000$, the function outputs a very large positive number.

Using symbols, we write: As $x \to -\infty$, $f(x) \to \infty$.

The end behavior can be seen on the graph of the function.

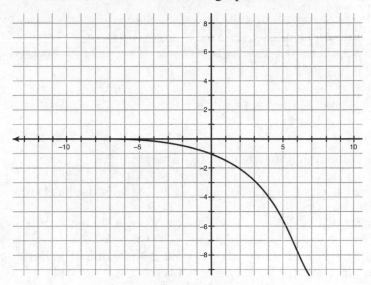

For large positive input values, this function outputs a large negative number. For large negative input values, this function outputs a number close to zero.

The end behavior of this function can be described:

$$\text{As } x \to \infty, f(x) \to -\infty$$

$$\text{As } x \to -\infty, f(x) \to 0$$

Check Your Understanding of Section 7.1

A. Multiple-Choice

1. If $f(x) = 2x + 3$ and $g(x) = 3x - 4$, what is $f(g(5))$?
   (1) 35   (2) 30   (3) 25   (4) 20

2. If $f(x) = 2x + 5$ and $g(x) = \dfrac{x - 5}{2}$, what is $g(f(7))$?
   (1) 6   (2) 7   (3) 8   (4) 9

3. Which of the following could be $f(x)$ and $g(x)$ if $f(g(x)) = (3x - 2)^2 + 1$?
   (1) $f(x) = x^2 + 1$, $g(x) = 3x - 2$
   (2) $f(x) = 3x - 2$, $g(x) = x^2 + 1$
   (3) $f(x) = x^2 - 1$, $g(x) = 3x - 2$
   (4) $f(x) = x^2 + 1$, $g(x) = 3x + 2$

4. If $f(g(x)) = 5$, what is $g(f(x))$?

   (1) 5   (2) –5   (3) $\dfrac{1}{5}$

   (4) Not enough information to answer

5. If $f(x) = x + 4$, $g(x) = 3x$, and $h(x) = x^2$, what is $f(h(g(x)))$?
   (1) $(3x + 4)^2$ (3) $3(x^2 + 4)$
   (2) $3x^2 + 4$ (4) $(3x)^2 + 4$

6. If $f(5) = 11$ and $g(x) = x^2 + 5$, what is the value of $g(f(5))$?
   (1) 126   (2) 16   (3) 11   (4) 5

7. If $f(x) = 3x - 7$, what is $f(4x)$?
   (1) $12x - 7$ (2) $12x^2 - 7$ (3) $12x - 28$ (4) $12x^2 - 28$

8. If $g(x) = 3x^2 - 2$, what is $g(f(x))$?
   (1) $2[f(x)]^2 + 3$ (3) $3[f(x)]^2 - 2$
   (2) $2[f(x)]^2 - 3$ (4) $3[f(x)]^2 + 2$

9. If $f(g(x)) = 2(x^2 + 1) - 5$ and $f(x) = 2x - 5$, what is $g(x)$?
   (1) $x + 1$ (2) $2x + 5$ (3) $5x - 2$ (4) $x^2 + 1$

10. If $f(x) = x^2 + 1$, what is the value of $f(f(f(f(f(0)))))$?
   (1) 2   (2) 5   (3) 26   (4) 677

B. *Show how you arrived at your answers.*

1. $f(x) = 3x - 2$ and $g(x) = 2x + 3$. Griffin says $f(g(3)) = 25$. Ruben says $f(g(3)) = 17$. Who is correct?

2. If $f(x) = x - 4$, $g(x) = x^2$, and $h(x) = 3x$, what is $h(g(f(x)))$ in simplified form?

3. Given the two graphs of $y = f(x)$ and $y = g(x)$, what is the value of $f(g(2))$?

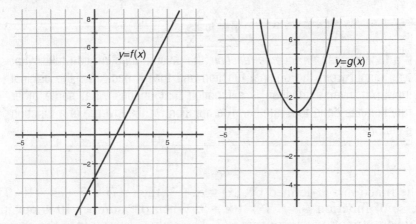

4. If $f(x) = x + 2$ and the graph of $y = g(x)$ is shown below, what does the graph of $f(g(x))$ look like?

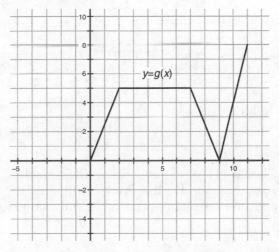

5. If $f(x) = x - 2$ and $g(x) = x^2$, for what value(s) of x does $f(g(x)) = g(f(x))$?

255

7.2 INVERSE FUNCTIONS

If the function f is defined by $f(x) = x + 3$, then $f(7) = 7 + 3 = 10$. The inverse of function f is $f^{-1}(x) = x - 3$. When the number 10 is put into the inverse function, the output is 7. This value was put into the original function, f, to get the number 10: $f^{-1}(10) = 10 - 3 = 7$. Whatever the function f does to a number, the function f^{-1} "undoes" to that number.

Example 1

If $g(6) = 19$, what is the value of $g^{-1}(19)$?

(1) 6

(2) $\dfrac{1}{6}$

(3) 19

(4) −19

Solution: Choice (1) is correct. Since the function g turned the number 6 into the number 19, the inverse will turn the 19 back into the 6. With the given information, this is the only value for g^{-1} that can be determined.

Some Basic Inverse Functions

If a function has one operation happening to the input value, the inverse will have the opposite operation happening to its input variable. Some examples of basic inverse functions are shown in the table.

$f(x)$	$f^{-1}(x)$
$x + 2$	$x - 2$
$x - 2$	$x + 2$
$2x$	$\dfrac{x}{2}$
$\dfrac{x}{2}$	$2x$
x^2 (for $x \geq 0$)	$\sqrt{x}$
$\sqrt{x}$	x^2 (for $x \geq 0$)
x	x
$-x$	$-x$
$\dfrac{1}{x}$	$\dfrac{1}{x}$

For any of these function/inverse pairs, you can verify that the composition

$$f(f^{-1}(9)) = 9$$

Determining the Inverse Function of a Linear Function

A linear function of the form $f(x) = ax + b$, such as $f(x) = 2x + 3$, has an inverse function that is also a linear function. The process for finding the inverse function is to first rewrite the function but with the $f(x)$ replaced with an x and the x replaced with an $f^{-1}(x)$.

$$x = 2f^{-1}(x) + 3$$

Then solve for $f^{-1}(x)$. In this case, subtract 3 from both sides of the equation and divide both sides of the equation by 2.

$$x = 2f^{-1}(x) + 3$$
$$\underline{-3 = -3}$$
$$x - 3 = 2f^{-1}(x)$$
$$\frac{x-3}{2} = \frac{2f^{-1}(x)}{2}$$
$$\frac{x-3}{2} = f^{-1}(x)$$

If you pick a number for x like 5, you can see that $f(5) = 2 \cdot 5 + 3 = 10 + 13$ and that $f^{-1}(13) = \dfrac{13-3}{2} = \dfrac{10}{2} = 5$. So this inverse function undoes what the function did to the number 5.

Example 2

What is the inverse of the function $f(x) = 3x - 5$?

(1) $f^{-1}(x) = \dfrac{x}{3} + 5$

(2) $f^{-1}(x) = \dfrac{x}{3} - 5$

(3) $f^{-1}(x) = \dfrac{x+5}{3}$

(4) $f^{-1}(x) = \dfrac{x-5}{3}$

Solution: Choice (3) is correct. Replace the $f(x)$ with an x and replace the x with an $f^{-1}(x)$. Solve the equation $x = 3f^{-1}(x) - 5$ for $f^{-1}(x)$ by adding 5 to both sides of the equation and then dividing both sides of the equation by 3.

$$x = 3f^{-1}(x) - 5$$
$$\underline{+5 = +5}$$
$$x + 5 = 3f^{-1}(x)$$
$$\frac{x+5}{3} = \frac{3f^{-1}(x)}{3}$$
$$\frac{x+5}{3} = f^{-1}(x)$$

Answering Questions About the Inverse Without Determining the Inverse

A trick question that appears on many Regents exams is one like the following.

If $f(x) = 5x - 2$, what is $f^{-1}(28)$?

The long way to do this question is to use the inverse process to find that the inverse function is $f^{-1}(x) = \dfrac{x+2}{5}$ and then calculate $f^{-1}(28) = \dfrac{28+2}{5} = \dfrac{30}{5} = 6.$

Since $f^{-1}(28)$ is asking what input value for x will output the number 28, a shorter way to do this is to simply solve the equation $28 = 5x - 2$.

$$28 = 5x - 2$$
$$\underline{+2 = +2}$$
$$30 = 5x$$
$$\frac{30}{5} = \frac{5x}{5}$$
$$6 = x$$

Example 3

If $f(x) = x^3 + 4$, what is the value of $f^{-1}(12)$?

Solution: Solve the equation $12 = x^3 + 4$.

$$12 = x^3 + 4$$
$$\underline{-4 = -4}$$
$$8 = x^3$$
$$\sqrt[3]{8} = \sqrt[3]{x^3}$$
$$2 = x$$

Graphs of Inverse Functions

If, for some function f, $f(3) = 9$, then the point $(3, 9)$ will be on the graph for that function. If $f(3) = 9$, then $f^{-1}(9) = 3$. So the point $(9, 3)$ will be on the graph for the inverse function. In general, if (x, y) is a point on the graph of the original function, then (y, x) will be a point on the graph of the inverse function. In terms of transformations, the point $(9, 3)$ is the reflection over the line $y = x$ of the point (3, 9). When this transformation is done to every point in the graph of the function, the graph of the inverse will be a reflection of the entire graph of the original function over the line $y = x$.

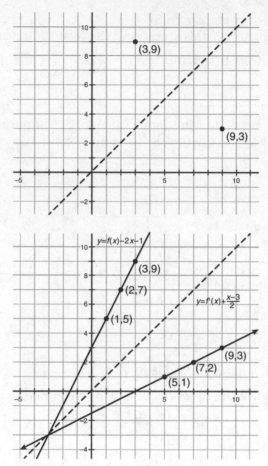

Some more graphs of functions and their inverses are shown in the following table.

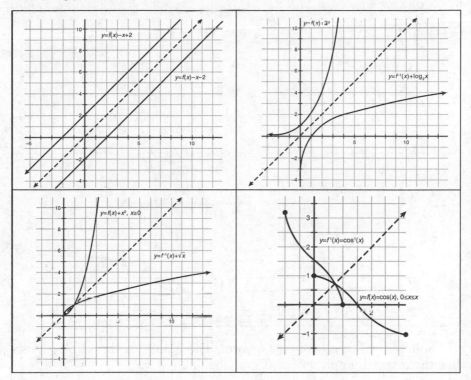

Example 4

Which choice is the graph of the inverse of the function whose graph is below?

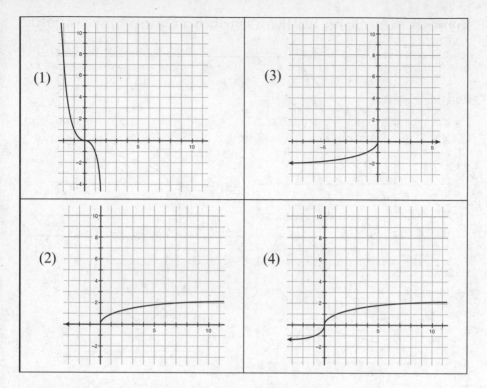

Solution: Choice (4) is the answer. The graph for this choice looks like the original graph reflected over the line $y = x$.

Inverse Functions on the Graphing Calculator

The TI-84 and TI-Nspire can easily graph inverse functions. To graph $y = f(x)$ and $y = f^{-1}(x)$ for the function $f(x) = 2x + 3$, follow these steps.

For the TI-84:

Press [Y1] and enter $2x + 3$ after Y1=. Graph this function by pressing [ZOOM] [6]. To graph the inverse, press [2ND] [PRGM] [8] for DrawInv. Then press [VARS] [RIGHT] [1] [1] and [ENTER]

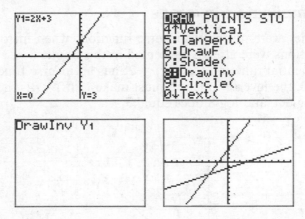

For the TI-Nspire:

From the homescreen, press [B] for the Graph Scratchpad. Press [tab] and enter $2x + 3$ after $f1(x) =$. Then press enter to see the graph of $f(x) = 2x + 3$. Press and hold [ctrl]. While keeping the [ctrl] key down, press [menu]. Then press [4] for Text. Type "$x = f1(y)$" and then press [enter]. Move the cursor onto this text. Press and hold [ctrl] and then [click]. Drag the text onto the x-axis, and it will graph the inverse.

More Complicated Inverses

Although they were not called inverse functions when introduced, two inverse functions were already presented in this book.

An exponential function like $f(x) = 2^x$ has an inverse function. Since $f(3) = 2^3 = 8$, the inverse function must make $f^{-1}(8) = 3$. The inverse of f $f^{-1}(x) = \log_2(x)$, and $\log_2 8$ does equal 3.

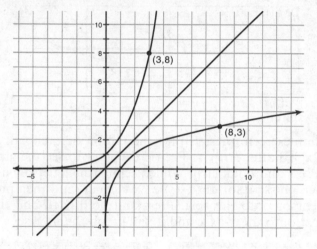

The trig function $f(x) = \cos(x)$, $0° \leq x < 180°$, has the inverse function $f^{-1}(x) = \cos^{-1}(x)$. So $\cos(0°) = 1$ and $\cos^{-1}(1) = 0°$.

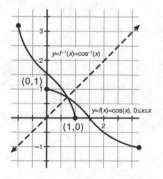

Check Your Understanding of Section 7.2

A. Multiple-Choice

1. If $g(x)$ is the inverse of $f(x)$, which of the following is *not* true?
   (1) $f(g(x)) = g(f(x))$
   (2) $g(x) = \dfrac{1}{f(x)}$
   (3) $f(g(x)) = x$
   (4) $g(f(x)) = x$

2. If $g(x)$ is the inverse of $f(x)$, what is the value of $f(g(9))$?
   (1) 3
   (2) $\dfrac{1}{9}$
   (3) -9
   (4) 9

3. What is the inverse of $f(x) = x - 3$?
   (1) $f^{-1}(x) = x + 3$
   (2) $f^{-1}(x) = \dfrac{1}{x+3}$
   (3) $f^{-1}(x) = x - 3$
   (4) $f^{-1}(x) = \dfrac{1}{x-3}$

4. What is the inverse of $f(x) = 4x$?
   (1) $f^{-1}(x) = \dfrac{x}{4}$
   (2) $f^{-1}(x) = 4x$
   (3) $f^{-1}(x) = \dfrac{1}{4x}$
   (4) $f^{-1}(x) = -4x$

5. What is the inverse of $f(x) = 2x - 7$?
   (1) $f^{-1}(x) = 2x + 7$
   (2) $f^{-1}(x) = \dfrac{1}{2x-7}$
   (3) $f^{-1}(x) = \dfrac{x+7}{2}$
   (4) $f^{-1}(x) = \dfrac{x}{2} + 7$

6. If $f(3) = 11$, what must also be true?
   (1) $f^{-1}(11) = 3$
   (2) $f^{-1}(3) = 11$
   (3) $f^{-1}(3) = \dfrac{1}{11}$
   (4) $f^{-1}(3) = -11$

7. If $f^{-1}(x) = 3x + 2$, what is $f(x)$?
   (1) $f(x) = \dfrac{1}{3x+2}$
   (2) $f(x) = \dfrac{x-2}{3}$
   (3) $f(x) = \dfrac{x}{3} - 2$
   (4) $f(x) = -3x - 2$

8. If the graph of $y = f(x)$ is shown below, what is the graph of $y = f^{-1}(x)$?

(1)

(2)

(3)

(4)

9. What is the inverse of $f(x) = 3^x$?

(1) $f^{-1}(x) = x^3$ (3) $f^{-1}(x) = 3^{\,x}$

(2) $f^{-1}(x) = \dfrac{1}{3^x}$ (4) $f^{-1}(x) = \log_3 x$

10. What is the inverse of $f(x) = \sin(x)$, $-90° \le x \le 90°$?

(1) $f^{-1}(x) = \dfrac{1}{\sin(x)}$ (3) $f^{-1}(x) = \sin^{-1}(x)$

(2) $f^{-1}(x) = \csc(x)$ (4) $f^{-1}(x) = \cos(x)$

B. *Show how you arrived at your answers.*

1. If $f(x) = 5x + 4$, what is $f^{-1}(34)$?

2. If the graph of $y = f(x)$ is shown below, sketch on the same axes the graph of $y = f^{-1}(x)$.

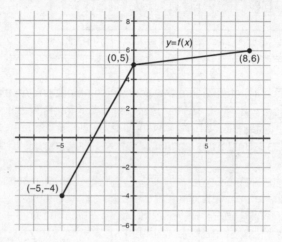

3. If $f(x) = 7x + 3$, what is $f^{-1}(x)$?

4. If $f(x) = \sqrt[3]{5x+19}$, what is $f^{-1}(f(x))$?

5. If $f(x) = 2x$ and $g(x) = x + 7$, show how you could solve $2x + 7 = 17$ by using the inverse functions $f^{-1}(x)$ and $g^{-1}(x)$.

SEQUENCES

8.1   SEQUENCES AND SERIES

KEY IDEAS

A *sequence* is a list of numbers that follows some pattern. An example of a sequence is $a = 1, 4, 9, 16, \ldots$. The individual terms of the sequence are called a_1, a_2, a_3, etc., where the small subscript is the term's position on the list. The terms of the sequence can be defined with two types of formulas. An *explicit formula* describes how to calculate the term by knowing the term's position on the list. A *recursive formula* describes how to calculate the term by knowing the previous terms on the list. The explicit formula is generally preferred but often more difficult to find.

Explicit Formula for a Sequence

An explicit formula is a type of function into which only positive integers can be input. Sometimes the explicit formula notation resembles function notation like $a(n) = 2n + 3$. An alternative notation is with a subscript, like $a_n = 2n + 3$. By substituting the values 1, 2, 3, and 4 for n, the first four terms of the sequence can be determined.

$$a_1 = 2 \cdot 1 + 3 = 5$$
$$a_2 = 2 \cdot 2 + 3 = 7$$
$$a_3 = 2 \cdot 3 + 3 = 9$$
$$a_4 = 2 \cdot 4 + 3 = 11$$

The sequence is $a = 5, 7, 9, 11, \ldots$.

MATH FACTS

An *infinite sequence* is one that has no end. The ellipses (…) at the end indicates that the sequence is infinite. A *finite sequence* does have a final term. The sequence 5, 7, 9, 11 is a finite sequence, while the sequence 5, 7, 9, 11, … is an infinite sequence.

Example 1

What are the first three terms of the sequence defined by equation $a_n = 5 + 3(n - 1)$?

   Solution: Substitute 1, 2, and 3 for n in the formula to get $a_1 = 5 + 3(1 - 1) = 5 + 3(0) = 5$, $a_2 = 5 + 3(2 - 1) = 5 + 3(1) = 8$, $a_3 = 5 + 3(3 - 1) = 5 + 3(2) = 11$.

Recursive Formula for a Sequence

A recursive formula has two parts. First there is the *base case*, where the first value (or the first few values) of the sequence are given. Second there is the *recursive definition*, where a rule for calculating more terms is described in terms of the previous terms.

   For example:

$$a_1 = 5$$

$$a_n = 2 + a_{n-1} \text{ for } n \geq 2$$

To get a_1, there is no calculation needed since a_1 is given as 5 in the base case. For a_2, substitute 2 for n into the recursive part of the definition.

$$a_2 = 2 + a_{2-1} = 2 + a_1 = 2 + 5 = 7$$

Notice that the value of a_1 needs to be known to do this step.

For $n = 3$, $a_3 = 2 + a_{3-1} = 2 + a_2 = 2 + 7 = 9$.

The first three terms are 5, 7, and 9.

   Notice that this recursive formula generates the same sequence as does the explicit formula $a_n = 2n + 3$.

Example 2

What is the value of a_3 for the sequence defined by

$$a_1 = 3$$

$$a_n = n \cdot a_{n-1}, \text{ for } n \geq 2?$$

   Solution: $a_1 = 3$ is given.
Substitute 2 for n into the recursive part of the definition.

$$a_2 = 2 \cdot a_{2-1} = 2 \cdot a_1 = 2 \cdot 3 = 6$$

Substitute 3 for n into the recursive part of the definition.

$$a_3 = 3 \cdot a_{3-1} = 3 \cdot a_2 = 3 \cdot 6 = 18$$

Example 3

What is the a_7 term of the sequence defined recursively as the following?

$$a_1 = x + 2$$

$$a_n = x \cdot a_{n-1}$$

Solution:

$$a_2 = x \cdot a_1 = x(x + 2) = x^2 + 2x$$

$$a_3 = x \cdot a_2 = x(x^2 + 2x) = x^3 + 2x^2$$

$$a_4 = x \cdot a_3 = x(x^3 + 2x^2) = x^4 + 2x^3$$

Each term becomes $x^n + 2x^{n-1}$. So a_7 will be $x^7 + 2x^6$.

Arithmetic and Geometric Sequences

A sequence like 5, 7, 9, 11, ... is known as an *arithmetic sequence* since each term is equal to 2 more than the previous term. The number 2, in this case, is known as the *common difference*, usually denoted by the variable d.

A sequence like 5, 10, 20, 40, ... is known as a *geometric sequence* since each term is equal to 2 times the previous term. The number 2, in this case, is known as the *common ratio*, usually denoted by the variable r.

The two formulas for the nth term of an arithmetic or of a geometric sequence are provided on the reference sheet given to you in the Regents booklet.

Arithmetic Sequence	**Geometric Sequence**
$a_n = a_1 + (n - 1)d$	$a_n = a_1 r^{n-1}$

Example 4

Find the explicit formula for the a_n term of sequence 7, 11, 15, 19, ...

Solution: Since each term is equal to the previous term plus 4, this is an arithmetic sequence: $a_1 = 7$ and $d = 4$. The formula is $a_n = 7 + (n - 1)4$.

Example 5

Find the explicit formula for the a_n term of sequence 7, 28, 112, 448, ...

Solution: Since each term is equal to the previous term times 4, this is a geometric sequence: $a_1 = 7$ and $r = 4$. The formula is $a_n = 7 \cdot 4^{n-1}$.

Example 6

What is the 50th term of the sequence 7, 11, 15, 19, ...?

Solution: The explicit formula is $a_n = 7 + 4(n - 1)$. Substitute 50 for n to get the answer.

$$a_{50} = 7 + 4(50 - 1) = 7 + 4(49) = 203$$

Example 7

The sequence $a = 3, 11, 19, 27, \ldots$ will eventually reach the number 251. If $a_n = 251$, what is the value of n?

Solution: The formula is $a_n = 3 + 8(n - 1)$. If $a_n = 251$, the equation becomes

$$251 = 3 + 8(n - 1)$$

One way to solve this is first to subtract 3 from both sides of the equation.

$$251 = 3 + 8(n - 1)$$
$$\underline{-3 = -3}$$
$$248 = 8(n - 1)$$

Now divide both sides of the equation by 8.

$$\frac{248}{8} = \frac{8(n-1)}{8}$$

$$31 = n - 1$$

Then add 1 to both sides of the equation.

$$31 = n - 1$$
$$\underline{+1 = +1}$$
$$32 = n$$
$$a_{32} = 251$$

Finite Arithmetic and Geometric Series

A *finite series* is like a finite sequence except all the terms are added together. An example of a finite arithmetic series is $5 + 7 + 9 + 11 + 13 + 15$. An example of a finite geometric series is $5 + 10 + 20 + 40 + 80 + 160$.

The formula for calculating the sum of a finite geometric series is given in the reference sheet in the Regents booklet. The formula is shown here.

$$S_n = \frac{a_1 - a_1 r^n}{1 - r}$$

To use this formula to calculate $5 + 10 + 20 + 40 + 80 + 160$, substitute 6 for n (since there are 6 terms), 5 for a_1, and 2 for r.

$$S_6 = \frac{5 - 5 \cdot 2^6}{1 - 2} = \frac{5 - 5 \cdot 64}{-1} = \frac{-315}{-1} = 315$$

The formula for calculating the sum of a finite arithmetic series is not given in the reference sheet in the Regents booklet. That formula is shown here.

$$S_n = \frac{n(a_1 + a_n)}{2}$$

Example 8

What is the sum of the first 100 positive integers in the following series?
$1 + 2 + 3 + \ldots + 98 + 99 + 100$

Solution: Substitute 100 for n, 1 for a_1, and 100 for a_n.

$$S_n = \frac{100(1 + 100)}{2} = \frac{100 \cdot 101}{2} = \frac{10,100}{2} = 5,050$$

Check Your Understanding of Section 8.1

A. Multiple-Choice

1. What are the first four terms of the sequence defined by $a_1 = 5$, $a_n = 3 + a_{n-1}$ for $n > 1$?
   (1) 5, 8, 11, 14   (3) 5, 2, –1, –4
   (2) 5, 15, 45, 135   (4) 5, 3, 1, –1

2. The sequence 3, 12, 48, 192, ... can be defined by which of the following?
   (1) $a_1 = 3$, $a_n = 8 + a_{n-1}$ for $n > 1$
   (2) $a_1 = 3$, $a_n = 4a_{n-1}$ for $n > 1$
   (3) $a_1 = 3$, $a_n = 12a_{n-1}$ for $n > 1$
   (4) $a_1 = 3$, $a_n = 12 + a_{n-1}$ for $n > 1$

3. What is the 20th term of the sequence 4, 11, 18, 25, ...?
   (1) 130   (2) 137   (3) 144   (4) 151

4. What position is the number 247 in the sequence 4, 13, 22, 31, ...?
   (1) 26   (2) 27   (3) 28   (4) 29

5. What is the 10th term of the sequence 4, 20, 100, 500, ...?
   (1) 7,812,500   (3) 195,312,500
   (2) 39,062,500   (4) 976,562,500

6. What is the sum of the series $5 + 13 + 21 + 29 + ... + 317$?
   (1) 6,380   (2) 6,400   (3) 6,420   (4) 6,440

7. What is the sum of the series $2 + 2 \cdot 3 + 2 \cdot 3^2 + 2 \cdot 3^3 + ... + 2 \cdot 3^{15}$?
   (1) $\dfrac{2(1-3^{15})}{1-3}$ (3) $\dfrac{2(1+3^{16})}{1+3}$

   (2) $\dfrac{2(1+3^{15})}{1+3}$ (4) $\dfrac{2(1-3^{16})}{1-3}$

8. A series is defined as $d_1 = 0$, $d_2 = 1$, $d_n = (n-1)(d_{n-2} + d_{n-1})$ for $n > 2$. What is the value of d_5?
   (1) 2   (2) 9   (3) 44   (4) 265

---

9. What is the sum of the first 20 terms of $2 - 6 + 18 - 54 + 162 - 486 + ...$?
   (1) $-1,743,392,000$
   (3) $-1,743,392,200$
   (2) $-1,743,392,100$
   (4) $-1,743,392,300$

10. $\dfrac{1}{2} + \dfrac{1}{4} + \dfrac{1}{8} + \dfrac{1}{16} + ... + \dfrac{1}{2^{20}}$ is closest to which of the following?
    (1) 0.99995   (2) 0.999995   (3) 0.9999995   (4) 1

B. *Show how you arrived at your answers.*

1. For 20 years, Journey puts $1,000 into the bank every January 1. The bank pays 3% interest every December 31. If she starts on January 1, 2000, the amount of money she will have on January 2, 2020 can be shown by the sequence:

   $1,000 + 1,000(1.05) + 1,000(1.05^2) + ... + 1,000(1.05)^{20}$. How much money is this?

2. What is greater: the 20th term of the sequence 10,000; 20,000; 30,000; 40,000; ... or the 20th term of the sequence 1, 2, 4, 8, ...?

3. The Fibonacci sequence is defined by $F_1 = 1$, $F_2 = 1$, $F_n = F_{n-1} + F_{n-2}$ for $n > 2$. What is F_{10}?

4. If $a_1 = 3$ and $a_n = n^2 \cdot a_{n-1} + n$, what is a_4?

5. 2,657,205 is what term number in the sequence 5, 15, 45, 135, ...?

PROBABILITY

9.1 SAMPLE SPACES

A *sample space* is a list of all the possible *outcomes* of some experiment. When it is possible to create an accurate sample space, calculating the *probability* of different types of outcomes can be found by counting. The probability of something happening can range from as low as 0% = 0.00 (if it is impossible) to as high as 100% = 1.00 (if it must happen). Most probabilities are somewhere between 0 and 1.

When a fair coin is flipped, there are two possible *outcomes*: heads and tails. Since each outcome is equally likely, the two outcomes can be written as a set called the *sample space*.

{Heads, Tails}

Anything you can try to calculate the probability of is known as an *event*. The sample space can be used to calculate the *probability* of different events that can happen with the coin.

To calculate the probability of the coin landing on heads, first count the number of possible outcomes in the sample space. Make that the denominator of your answer. Then count the number of *favorable* outcomes, meaning outcomes in the sample space that are heads. Make that the numerator of your answer. Since there are 2 possible outcomes and 1 outcome with heads, the probability of the coin landing on heads is $\frac{1}{2}$.

MATH FACTS

If there is a sample space, the probability of some event happening can be

calculated with the fraction $\dfrac{\text{Favorable outcomes in the sample space}}{\text{Total outcomes in the sample space}}$.

Below is a picture of a circle divided into 6 equal slices, numbered 1 though 6. A spinner is put into the middle and spun.

The sample space for one spin is {1, 2, 3, 4, 5, 6}.
With this sample space, various questions can be answered.

Basic Probability Questions

The most basic question is one like: What is the probability that the spinner will land on the number 4? This question is abbreviated as P(4).

Since there is one 4 in the sample space and there are 6 total numbers in the sample space, $P(4) = \dfrac{1}{6}$.

Probability Questions Involving the Word "Not"

If the question is to find the probability that the spinner will not land on the number 4, or P(not 4), count the number of outcomes in the sample space that are not 4. Since 5 of the numbers in the sample space are not the number 4, $P(\text{not } 4) = \dfrac{5}{6}$.

Probability Questions Involving the Word "And"

Questions involving the word "and" require you to analyze each element in the sample space to see if it meets two different conditions. Here's a typical question involving the spinner with 6 equal slices, numbered 1 through 6.

"On one spin, what is the probability that the spinner will land on a number that is greater than 3 and even, or P(greater than 3 and even)?"

The denominator of the solution is still 6. For the numerator, examine all 6 numbers in the sample space to see how many of them are both greater than 3 and also even.

- 1 is neither even nor greater than 3.
- 2 is even but is not greater than 3.
- 3 is neither even nor greater than 3.
- 4 is both even and greater than 3.
- 5 is greater than 3 but not even.
- 6 is both even and greater than 3.

Exactly two of the numbers, 4 and 6, satisfy both conditions. So the numerator of the probability fraction is 2, and the solution is

$$P(\text{even and greater than 3}) = \frac{2}{6} = \frac{1}{3}.$$

Probability Questions Involving the Word "Or"

Imagine there is a game where you spin the spinner and you win a prize if the spinner lands on an even number or on a number that is greater than 3. To find the probability of winning this game, go through the 6 outcomes to find how many of them are either even, greater than 3, or both.

Of the 6 numbers, the numbers 2, 4, 5, and 6 all are either even, greater than 3, or both. The probability of getting a number that is even or greater than 3 is $P(\text{even or greater than 3}) = \frac{4}{6} = \frac{2}{3}.$

Probability Questions Involving the Word "Given"

Imagine the spinner is spun. Somebody looks at it before you get a chance to and tells you a hint about what number it landed on, such as "It landed on an even number." Then a question is asked like, "What is the probability that it landed on a number greater than 3?" Had you not known about the hint, the solution would be $\frac{3}{6}$ since 3 of the 6 numbers (4, 5, and 6) are greater than 3. With the hint, though, you can get a more accurate answer.

This question could be described as "What is the probability of getting a number greater than 3 given that it landed on an even number?" or P(greater than 3 given even). You can instead use the "given" symbol, |, to write P(greater than 3|even).

To calculate the probability using the sample space, first use the given information to reduce the sample space, crossing out anything that cannot have happened. For this example, the sample space is now $\{\cancel{1}, 2, \cancel{3}, 4, \cancel{5}, 6\} = \{2, 4, 6\}$. Now the question can be answered using this new, more accurate sample space. There are 3 outcomes in this sample space, and 2 of them (4 and 6) are greater than 3. So $P(\text{greater than 3}|\text{even}) = \dfrac{2}{3}$.

Example 1

A fair coin is flipped, and a fair 6-sided die is rolled. The sample space has 12 outcomes {H1, H2, H3, H4, H5, H6, T1, T2, T3, T4, T5, T6}. Using this sample space, calculate the probability of getting:

A) An outcome with tails on the coin
B) An outcome with tails on the coin and 2 on the die
C) An outcome with tails on the coin or 2 on the die
D) An outcome with tails on the coin given that there is a 2 on the die
E) An outcome with a 2 on the die given that there is a tails on the coin

Solution:

A) Since 6 out of the 12 outcomes have a T in them, the probability is $\dfrac{6}{12} = \dfrac{1}{2}$.

B) Only one of the 12 outcomes has a T and a 2, so the probability is $\dfrac{1}{12}$.

C) The outcomes H2, T1, T2, T3, T4, T5, T6 each have a 2 or a T (T2 has both). This is 7 out of the 12 outcomes in the sample space, so the probability is $\dfrac{7}{12}$.

D) Since it is given that 2 is on the die, the sample space gets reduced to {H2, T2}. Only one of the two outcomes has a T, so the probability is $\dfrac{1}{2}$.

E) Since it is given that tails is on the coin, the sample space set gets reduced to {T1, T2, T3, T4, T5, T6}. Of the 6 possible outcomes in this reduced sample space, only one has a 2 on the die. So the probability is $\dfrac{1}{6}$.

Sample Spaces Represented as Tables

For certain surveys, the results can be represented in a table. This table can serve as a sample space for probability questions.

Imagine that 100 people are surveyed and asked two questions "Are you more than 14 years old?" and "Are you more than 5 feet tall?" The results of the survey can be shown with a table like this.

	Under 14 Years	Over 14 Years	Total
Under 5 feet	21	12	33
Over 5 feet	9	58	67
Total	30	70	100

A probability question that can be answered with this table is "If one person is taken from the 100 at random, what is the probability the person will be both under 14 years old and over 5 feet tall?"

The denominator of the fraction will be 100 since there are 100 people in total. For the numerator, the number of people who are both under 14 years old and over 5 feet tall, based on information in the table, is 21. So the solution is $\frac{21}{100} = .21$.

Example 2

Using the above table, what is the probability that a randomly chosen person will be under 14 years old?

Solution: The denominator of the solution is 100. The numerator is $21 + 9 = 30$ since there are 21 people under 14 years old who are under 5 feet tall and 30 people under 14 years old who are over 5 feet tall. The solution is $\frac{30}{100} = .30$.

Example 3

Using the above table, find the probability that a randomly chosen person is over 5 feet tall or over 14 years old.

Solution: The total number of people is 100, which will be the denominator of the solution. For the numerator, look at the four possibilities and see which count people who are either over 5 feet tall, over 14 years old, or both. Three of the four possibilities have at least one of those characteristics. The only one that doesn't is the people who are both under 14 years old and under 5 feet tall. Add 9 + 12 + 58 = 79, and make that the numerator. The solution is $\frac{79}{100} = .79$.

Example 4

Using the above table, find the probability of a randomly chosen person being over 5 feet tall given that the person is over 14 years old.

Solution: Because of the word "given," the denominator of the fraction is no longer going to be 100. Since it is known already that the person is over 14 years old, the 30 people under 14 years old are no longer relevant. The number of people over 14 years old is 70, and this will be the denominator of the solution. Of those 70 people, 58 of them are over 5 feet tall. So the solution is $\frac{58}{70} \approx .83$.

Sample Spaces Represented as Venn Diagrams

Another way to represent the data from the age/height survey is with a Venn diagram.

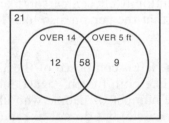

The two circles represent the people who are over 14 years old and the people who are over 5 feet tall. Just as there are four different categories of people in the table (both under 14 and under 5 feet, under 14 but over 5 feet, over 14 but under 5 feet, and both over 14 and over 5 feet), this Venn diagram has 4 regions (in neither circle, in the circle on the left but not in the circle on the right, in the circle on the right but not in the circle on the left, and in both circles).

Use the Venn diagram to answer a question like "If a person is chosen randomly from the 100 surveyed, what is the probability that the person is under 14 years old but over 5 feet tall?" The circle on the right represents the people who are over 5 feet tall. This is composed of the 58, which are the people who are both over 5 feet tall and over 14 years old, and of the 9, which are the people who are over 5 feet tall but not over

14 years old (since they are not inside the over 14 years old circle on the right). The solution is $\frac{9}{100} = .09$.

Example 5

The Venn diagram below is based on two survey questions "Do you like frozen yogurt?" and "Do you like mustard?"

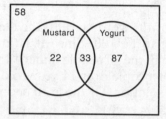

Use this Venn diagram to answer the following questions:

A) What is the probability that a person chosen randomly among those surveyed likes frozen yogurt but does not like mustard?
B) What is the probability that a person chosen randomly among those surveyed likes mustard?
C) Given that a randomly chosen person likes mustard, what is the probability that the person also likes frozen yogurt?

Solution:

A) The circle on the left is composed of the 33, which is the number of people who like both frozen yogurt and mustard, and of the 87, which is the number of people who like frozen yogurt (they are inside the Y circle) but do not like mustard

(they are outside the M circle). The solution is $\frac{87}{200} = .435$.

B) The M circle has 33 + 22 = 55. So the solution is $\frac{55}{200} = .275$.

C) Since it is given that the person likes mustard, the denominator of the fraction will not be the total surveyed but just those surveyed who like mustard. From part (b), this is 55. Since 33

of those 55 like frozen yogurt, the solution is $\frac{33}{55} = .60$.

Check Your Understanding of Section 9.1

A. Multiple-Choice

1. If a fair penny and a fair dime are flipped, the sample space of possible outcomes is {HH, HT, TH, TT}. What is the probability of getting tails on both coins?

   (1) $\dfrac{1}{2}$ (2) $\dfrac{1}{4}$ (3) $\dfrac{3}{4}$ (4) 1

2. A fair 6-sided die is rolled. The sample space is {1, 2, 3, 4, 5, 6}. What is the probability that the number that comes up is both even and greater than 4?

   (1) $\dfrac{1}{6}$ (2) $\dfrac{2}{6}$ (3) $\dfrac{3}{6}$ (4) $\dfrac{4}{6}$

3. A spinner with equal-size sections and with the numbers 1 to 8 on it is spun. The sample space is {1, 2, 3, 4, 5, 6, 7, 8}. What is the probability that the number the spinner lands on is either even or greater than 5?

   (1) $\dfrac{3}{8}$ (2) $\dfrac{4}{8}$ (3) $\dfrac{7}{8}$ (4) $\dfrac{5}{8}$

4. A fair coin is tossed, and a fair 6-sided die is rolled. The sample space of possible outcomes is {H1, H2, H3, H4, H5, H6, T1, T2, T3, T4, T5, T6}. If it is known that the die landed on a number greater than 4, what is the probability that the coin landed on heads?

   (1) $\dfrac{1}{2}$ (2) $\dfrac{1}{3}$ (3) $\dfrac{1}{4}$ (4) $\dfrac{1}{6}$

Questions 5, 6, and 7 use the following information:

40 people are surveyed about whether they like iOS or Android. The results are collected on this table.

	Like Android	Don't Like Android	Total
Like iOS	7	20	27
Don't like iOS	12	1	13
Total	19	21	40

5. If one of the 40 people is selected at random, what is the probability the person likes both iOS and Android?

(1) $\dfrac{39}{40}$ (2) $\dfrac{7}{40}$ (3) $\dfrac{46}{40}$ (4) $\dfrac{1}{40}$

6. What is the probability that a randomly selected person likes Android if it is known that the person likes iOS?

(1) $\dfrac{7}{27}$ (2) $\dfrac{19}{40}$ (3) $\dfrac{7}{40}$ (4) $\dfrac{19}{27}$

7. What is the probability that a randomly selected person likes either iOS or Android?

(1) $\dfrac{39}{40}$ (2) $\dfrac{37}{40}$ (3) $\dfrac{35}{40}$ (4) $\dfrac{32}{40}$

Questions 8, 9, and 10 use the following information:

50 students are surveyed about how they like the red and blue Barron's books. The results are collected on this Venn diagram.

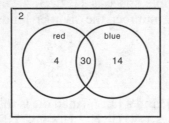

8. What is the probability that a randomly selected person likes both the red and the blue book?

(1) .30   (2) .40   (3) .50   (4) .60

9. What is the probability that a randomly chosen person will not like the blue book?

(1) .06   (2) .12   (3) .18   (4) .24

10. From the group of people who like the red book, a person is randomly chosen. What is the probability that the person also likes the blue book?

(1) .72   (2) .80   (3) .88   (4) .96

B. *Show how you arrived at your answers.*

1. A fair 6-sided die with the faces numbered 1 to 6 is rolled. A spinner with the letters A, B, and C, with equal chances of occurring, is spun. What is the sample space of possible outcomes?

2. A family with three children can be one of eight possibilities:

   {BBB, BBG, BGB, BGG, GBB, GBG, GGB, GGG}.

   What is the probability of the family having 2 boys and a girl in any order?

3. 80 students are asked two questions: (1) Do you like the Rangers? and (2) Do you like the Islanders? 40 people answered yes to question 1, and 55 people answered yes to question 2. 10 people answered no to both. How many people answered yes to both? Use this Venn diagram to assist you.

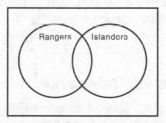

4. 200 students get to choose an ice cream cone. They can pick one flavor from the three choices: chocolate, vanilla, and strawberry. They can pick one topping, either sprinkles or peanuts. The results are collected on this table.

	Sprinkles	**Peanuts**	**Total**
Chocolate	40	50	90
Vanilla	25	5	30
Strawberry	65	15	80
Total	130	70	200

   a) What is the probability that a randomly selected person chose chocolate ice cream?
   b) What is the probability that a person who chose chocolate ice cream also chose peanuts?
   c) What is the probability that a person who chose sprinkles also chose strawberry ice cream?

285

5. A family has 4 children.
   a) Make a sample space of all the boy/girl combinations; there are 16 of them.
   b) What is more likely: having a 2/2 split (2 boys and 2 girls) or a 3/1 split (either 3 boys and 1 girl or 3 girls and 1 boy)?

9.2 CALCULATING PROBABILITIES INVOLVING INDEPENDENT EVENTS

△
KEY IDEAS
╱‾‾╲

Two events are *independent* if one happening (or not happening) has nothing to do with whether or not the other happens (or doesn't happen). When events are not independent, they are *dependent*. It is much simpler to calculate the probability when questions that include the words "or," "and," and "given" involve independent events.

Dependent Events vs. Independent Events

Many things in the real word depend on other things. If you ask someone, "Are you going to the beach next Saturday?" that person could say, "It depends." If you follow up with "Depends on what?" the person could respond, "On what the weather is like," "On whether or not my friend with a car is working," or all kinds of other possibilities.

If you ask the same person, "Is your birthday next Saturday?" he or she will not likely say, "It depends," since there isn't anything else that will make it more likely or less likely that the person's birthday is next Saturday. Next Saturday will be their birthday or it will not, regardless, for example, of what the weather is like.

The events "person will go to the beach on Saturday" and "the weather is nice on Saturday" are dependent events. The events "person's birthday is Saturday" and "the weather is nice on Saturday" are independent events.

The table shows some events and whether or not they qualify as dependent or as independent when combined.

Event A	Event B	Dependent or Independent
A coin is flipped	A die is rolled	Independent
What happens in the Yankees baseball game	What happens in the Rangers hockey game	Independent
Winning the lottery	Buying an airplane	Dependent
Passing the Algebra II Regents	Studying this book	Dependent
Rooting for the Mets	The Mets winning	Probably Independent, but maybe if you're at the game and cheering really hard, it could help a little!

MATH FACTS

It is sometimes not very clear whether two events are dependent or independent. Questions involving coin tosses, spinners, and dice are generally about independent events. For events involving human behavior, an argument can sometimes be made for either dependent or independent.

Example 1

Do you think these events are dependent or independent? Explain your reasoning.

Event *A*: The groundhog sees his shadow on Groundhog's Day.

Event *B*: There are six more weeks of cold weather.

Solution: As this is an opinion question, either independent or dependent is correct as long as your reasoning is clear. For independent, you could say that the weather does not in any way know whether the groundhog saw his shadow or not. So winter will come whenever it does, regardless. For dependent, you could say that maybe the groundhog seeing his shadow means that it is sunny on Groundhog's Day and that the sunny day is an indication that there will be more sunny days in the future so the warm weather will come sooner.

Probability Questions Involving the Word "And"

When a coin is flipped and a 6-sided die is rolled, the outcome of the coin flip is independent of the outcome of the die roll. The coin does not know (or care, for that matter) what happened with the die.

In Section 9.1, sample spaces were used to answer probability questions involving the word "and." With independent events, there is a shortcut for this. The shortcut is partly justified by the fact that when one fraction is multiplied by another fraction, the result will be a fraction that is smaller than either of the fractions. Likewise, the probability of two things happening is smaller than either one of them occurring individually.

When two events are independent, the probability of the first event and of the second event both happening is equal to the product of the probabilities of each happening separately.

===== **MATH FACTS** =====

If event A and event B are independent events, the probability of

$$P(A \text{ and } B) = P(A) \cdot P(B).$$

If A is the event "the coin lands on tails," then $P(A) = \dfrac{1}{2}$. If B is the event "the die shows a 2," then $P(B) = \dfrac{1}{6}$. (Small sample spaces {H, T} and {1, 2, 3, 4, 5, 6} could be used to determine these probabilities.) Since these are independent events, the probability that the coin lands on heads and that the die shows a 2 is $P(A \text{ and } B) = P(A) \cdot P(B) = \dfrac{1}{2} \cdot \dfrac{1}{6} = \dfrac{1}{12}$. This is the same answer as was obtained in Section 9.1 with the large sample space.

Example 2

If the probability of event A, that it will rain in Boston on Friday, is .6 and the probability of event B, that the Knicks will win the game they play in Los Angeles on Friday, is .3, what is the probability that it both rains in Boston on Friday and that the Knicks win the game they play in Los Angeles on Friday?

Solution: These are independent events. Whether or not it rains in Boston on Friday has no impact on whether or not the Knicks win in Los Angeles on Friday. (Some people think that all things affect each other in tiny cosmic ways, even events like these. However, for the Regents, these are implied to be independent events.)

$$P(A \text{ and } B) = P(A) \cdot P(B) = .6 \cdot .3 = .18$$

Probability Questions Involving the Word "Or"

For independent events, the probability of either A or B (or both) happening can be calculated with the formula $P(A \text{ or } B) = P(A) + P(B) - P(A \text{ and } B)$. Since the events are independent, $P(A \text{ and } B) = P(A) \cdot P(B)$. So the formula becomes

$$P(A \text{ or } B) = P(A) + P(B) - P(A) \cdot P(B).$$

MATH FACTS

If events A and B are independent, then $P(A \text{ or } B) = P(A) + P(B) - P(A) \cdot P(B)$

Example 3

If the probability of event A, that it will rain in Boston on Friday, is .6 and the probability of event B, that the Knicks will win the game they play in Los Angeles on Friday, is .3, what is the probability that it rains in Boston on Friday or that the Knicks win the game they play in Los Angeles on Friday (or both)?

Solution: Since they are independent events, $P(A \text{ or } B)$ can be calculated with the formula $P(A) + P(B) - P(A) \cdot P(B) = .6 + .3 - .6 \cdot .3 = .72$.

Probability Questions Involving the Word "Given"

For independent events A and B, the probability of A happening is not affected by whether or not B happened. So when A and B are independent, $P(A|B) = P(A)$.

Example 4

If the probability of event A, that it will rain in Boston on Friday, is .6 and the probability of event B, that the Knicks will win the game they play in Los Angeles on Friday, is .3, what is the probability that the Knicks win on Friday given that it will rain in Boston on Friday?

Solution: Since the events are independent, $P(A|B) = P(A) = .3$.

Check Your Understanding of Section 9.2

A. *Multiple-Choice*

1. Which two events are independent of one another?
   (1) A = It rains in Nevada on January 1, 1975.
   B = A New York resident goes to the movies on May 17, 2016.
   (2) A = It is Valentine's Day.
   B = Flower sales have increased.
   (3) A = A student eats a healthy breakfast.
   B = A student has energy at 10:00 A.M.
   (4) A = A teacher wins the lottery.
   B = The same teacher quits his job.

2. Which two events are *not* independent?
   (1) A = A child has brown hair.
   B = The child's biological parents both have brown hair.
   (2) A = A coin lands heads side up.
   B = A 6-sided die lands with the 6 face up.
   (3) A = A student's birthday is on May 1.
   B = Another student seated next to that student is born on May 1.
   (4) A = A person wears a blue shirt.
   B = The same person likes the TV show "Seinfeld."

3. A and B are independent events. If P(A) = .4 and P(B) = .5 what is P(A and B)?
   (1) .4   (2) .5   (3) .2   (4) .9

4. A and B are independent events. If P(A) = .4 and P(B) = .5 what is P(A or B)?
   (1) .7   (2) .9   (3) .5   (4) .45

Questions 5, 6, and 7 use the following information:

A fair coin is flipped, and a spinner with the letters A, B, C, D, and E, all equally likely, is spun.

5. What is the probability the coins will land heads and the spinner will land on D?
   (1) $\dfrac{1}{2}$ (2) $\dfrac{1}{5}$ (3) $\dfrac{1}{10}$ (4) $\dfrac{7}{10}$

6. What is the probability that the coin will land on heads or the spinner will land on D?

(1) $\frac{4}{10}$ (2) $\frac{5}{10}$ (3) $\frac{6}{10}$ (4) $\frac{7}{10}$

7. What is the probability that the coin will land on heads given that the spinner landed on D?

(1) $\frac{1}{10}$ (2) $\frac{3}{10}$ (3) $\frac{7}{10}$ (4) $\frac{1}{2}$

8. If the probability that the Rangers win the Stanley Cup is .05 and the probability that the Yankees win the World Series is .15, what is the probability that both will happen?

(1) .20   (2) .0075   (3) .05   (4) .15

9. Anne will go out to dinner with a group of friends if the restaurant is nearby or if it is a sushi restaurant. If P(nearby) = .7 and P(sushi) = .2 and these are independent events, what is the probability of Anne going out with the group?

(1) .76   (2) .83   (3) .90   (4) .14

10. *A* and *B* are independent events. If P(*A*) = .6 and P(*B*) = .9, what is P(*A|B*)?

(1) .3   (2) .6   (3) .9   (4) .54

B. *Show how you arrived at your answers.*

1. Write an event that would be dependent with the event *A* = "It rains this Saturday."

2. If *A* is the event "I eat pizza on Tuesday" and *B* is the event "I eat roast beef on Saturday" and if P(*A*) = .4 and P(*B*) = .3, what is the probability that I eat pizza on Tuesday and I do *not* eat roast beef on Saturday?

3. The probability of it raining on Saturday is 70%. The probability of Nayeli going to the movies on Saturday is 30%. Nayeli reasons that it will definitely either rain or she will go to the movies since .70 + .30 = 1. Is this reasoning correct?

4. *A* and *B* are independent events. P(*A*) = .6 and P(*B*) = .2. What is P(*B|A*)?

5. The Mets are playing the Royals in baseball. The probability that the Mets will win is .45. The probability that the Royals will win is .55. What is the probability that the Mets will win *and* that the Royals will win?

9.3 CALCULATING PROBABILITIES INVOLVING DEPENDENT EVENTS

===== KEY IDEAS =====

When events are dependent, rather than independent, calculating probabilities is more complicated. If in a problem different probabilities are already given, it is possible to use them to determine if certain events are independent or dependent.

Dependent events can be complicated. If the probability of Camden going to a party is .8 and the probability of Greyson going to a party is .7, what is the probability of them both going to the party?

Well, it depends. If these were independent events, the probability of both going to the party would be the product of $.8 \cdot .7 = .56$. However, what if Camden and Greyson are rivals? What if when Camden got there and saw Greyson, he would turn around and leave? Then the probability of both going is not .56 but 0.

A simpler example of dependent events is choosing a doughnut from a box of doughnuts for event *A*, eating it, and then choosing a second doughnut for event *B*. Suppose the box of 12 doughnuts has 3 glazed, 5 jelly filled, 3 powdered sugar, and 1 custard filled.

Probability Questions Involving the Word "Given"

In the doughnut example, it is possible to calculate the probability that the second doughnut was glazed given that the first doughnut is glazed. If the first doughnut was glazed, after it is eaten there are only 11 doughnuts left in the box. Of those 11, only 2 are glazed. So the probability of the first doughnut being glazed is $P(A) = \frac{3}{12} = \frac{1}{4}$. However, the probability of the second doughnut being glazed given that the first doughnut was glazed is $P(B|A) = \frac{2}{11}$.

Example 1

If event A is "the first doughnut is custard filled" and event B is "the second doughnut is custard filled," what is the probability $P(B|A)$?

Solution: The first doughnut is custard filled. After it is eaten, there are no custard-filled doughnuts left. So it is impossible to get a custard filled for the second doughnut, $P(B|A) = 0$.

One Way of Determining if Events Are Independent or Dependent

If the probability of an event is known and the probability of that event given that another event happened is also known, it is possible to determine whether or not the two events are dependent events.

===== **MATH FACTS** =====

For dependent events, A and B, $P(A|B) \neq P(A)$. In other words, if B has happened, the probability of A happening is different than it would have been if B had not happened.

Example 2

If the probability of Joshua being late is .3 and the probability of Joshua being late given that he missed the bus is .9, what can be said about the events "Joshua is late to school" and "Joshua missed the bus"?

(1) They are independent events.

(2) They are dependent events.

(3) There is not enough information to determine if they are independent or dependent events.

(4) Joshua needs to wake up earlier.

Solution: Choice (2) is correct. Since P(Joshua is late) $\neq$ P(Joshua is late given that Joshua missed the bus), these are dependent events. If these probabilities were equal, the events would be independent.

Probability Questions Involving the Word "And"

With dependent events A and B, the probability of A and B both happening is not just $P(A) \cdot P(B)$ as it was with independent events.

With the box of doughnuts from before, the probability of getting a glazed doughnut as the first doughnut and getting a glazed doughnut as the second doughnut is $\frac{3}{12}$ (the probability of getting a glazed doughnut first) times $\frac{2}{11}$ (the probability of getting a glazed doughnut second given that the first one was glazed).

P(glazed doughnut first and glazed doughnut second)

$$= \frac{3}{12} \cdot \frac{2}{11} = \frac{6}{132} = .05$$

===== **MATH FACTS** =====

For dependent events A and B, P(A and B) = P(A) · P($B|A$). When the events are independent, this formula also works since for independent events P($B|A$) = P(B). So P(A) · P($B|A$) = P(A) · P(B), which is the formula for P(A and B) when A and B are independent.

Another Way of Determining if Events Are Independent or Dependent

If A and B are independent events, P(A and B) = P(A) · P(B). If they are not independent events, then P(A and B) ≠ P(A) · P(B). If the values of P(A), P(B), and P(A and B) are known, comparing P(A and B) to P(A) · P(B) will quickly reveal if the two events are dependent or independent.

Example 3

The probability of Genevieve having pizza for lunch on Wednesday is .4. The probability of her having pizza for dinner on Wednesday is .3. If the probability of Genevieve having pizza for both lunch and for dinner on Wednesday is .05, what can be concluded about event A, Genevieve will have pizza for lunch on Wednesday, and about event B, Genevieve will have pizza for dinner on Wednesday?

(1) The events are independent since P(A and B) = P(A) · P(B).

(2) The events are dependent since P($A|B$) ≠ P(A).

(3) The events are dependent since P(A and B) ≠ P(A) · P(B).

(4) The events are independent since P($A|B$) = P(A).

Solution: Since .3 · .4 ≠ .05, P(A and B) ≠ P(A) · P(B). This means A and B are dependent events. Choice (3) is correct. Although choice (2) also concludes they are dependent events, the value of P($A|B$) is not known, so it can't be compared to P(A).

Dependent or Independent Events with Tables or Venn Diagrams

In Section 9.1 the results of a survey were presented in this table.

	Under 14 Years	Over 14 Years	Total
Under 5 feet	21	12	33
Over 5 feet	9	58	67
Total	30	70	100

Using the topics from this section, it can be determined whether or not the event A, "The person is over 14 years old" and event B, "The person is over 5 feet tall" are dependent or independent.

One way to check is to see if $P(A) = P(A|B)$.

$P(A) = \dfrac{70}{100} = .7$. However, $P(A|B) = \dfrac{58}{67} = .87$. Since these are not equal, the events are not independent.

A second way to check is to see if $P(A \text{ and } B) = P(A) \cdot P(B)$.

$P(A \text{ and } B) = \dfrac{58}{100} = .58$, $P(A) = \dfrac{70}{100} = .7$, and $P(B) = \dfrac{67}{100} = .67$. Since $P(A) \cdot P(B) = .469$, which does not equal $P(A \text{ and } B) = .58$, the events are not independent.

These two tests can also be applied to a Venn diagram.

Example 4

The Venn diagram below is based on two survey questions, "Do you like frozen yogurt?" and "Do you like mustard?"

Based on the information in this Venn diagram, determine if event Y, "The person likes frozen yogurt" and event M, "The person likes mustard" are independent or dependent.

Solution: One way to check is to see if $P(Y \text{ and } M) = P(Y) \cdot P(M)$. For this example

$$P(Y \text{ and } M) = \frac{33}{200} = .165. \quad P(Y) = \frac{120}{200} = .60. \quad P(M) = \frac{55}{200} = .275.$$

Since $P(Y) \cdot P(M) = .60 \cdot .275 = .165$, these are independent events.

The other way this could also be checked is by comparing $P(Y) = \frac{120}{200} = .6$ to $P(Y|M) = \frac{33}{55} = .6$. Since $P(Y) = P(Y|M)$, they are independent events.

Calculating the Probability of *A* Given *B*

The formula $P(A \text{ and } B) = P(A) \cdot P(B|A)$ can be rearranged into a formula that enables you to calculate $P(B|A)$ if the right information is known.

MATH FACTS

For any events *A* and *B*, $P(B|A) = \dfrac{P(A \text{ and } B)}{P(A)}$. For independent events,

this becomes $P(B|A) = \dfrac{P(A \text{ and } B)}{P(A)} = \dfrac{P(A) \cdot P(B)}{P(A)} = P(B)$.

To use this formula, very specific information must be provided in the question.

Example 5

If the $P(A \text{ and } B)$ is .4 and $P(A)$ is .7, what is $P(B|A)$?

Solution: $P(B|A) = \dfrac{P(A \text{ and } B)}{P(A)} = \dfrac{.4}{.7} \approx .57$

Example 6

Event *B* is "patient has hepatitis." Event *A* is "patient tests positive for hepatitis." If $P(A) = .01098$ and $P(A \text{ and } B) = .00099$, what is the probability that the patient has hepatitis given that the person tests positive for hepatitis?

Solution: $P(B|A) = \dfrac{P(A \text{ and } B)}{P(A)} = \dfrac{.00099}{.01098} \approx .09 = 9\%$

Check Your Understanding of Section 9.3

A. Multiple-Choice

1. If $P(A) = .4$ and $P(A|B) = .5$, what conclusion can be made?
   (1) $P(B) = .6$
   (2) A and B are independent events.
   (3) A and B are not independent events.
   (4) No conclusion can be made.

2. If $P(A) = .7$ and if A and B are independent events, what must be true?
   (1) $P(B) \neq .7$ (3) $P(B|A) = .7$
   (2) $P(A|B) = .7$ (4) No conclusion can be made.

3. If $P(A) = .4$, $P(B) = .7$, and $P(B|A) = .9$, what is $P(A$ and $B)$?
   (1) 1.3   (2) .28   (3) .36   (4) .63

4. If $P(A) = .3$ and $P(A$ and $B) = .12$, what is $P(B|A)$?
   (1) .2   (3) .4
   (2) .3   (4) Not enough information.

5. A dish has 20 blue candies and 30 yellow candies. One candy is chosen at random and eaten. Then another one is chosen at random and eaten. What is the probability that both candies were blue?
   (1) $\dfrac{20}{50} + \dfrac{19}{49}$ (2) $\dfrac{20}{50} \cdot \dfrac{20}{50}$ (3) $\dfrac{20}{50} \cdot \dfrac{19}{50}$ (4) $\dfrac{20}{50} \cdot \dfrac{19}{49}$

6. If $P(A) = .4$, $P(B) = .6$, and $P(A$ and $B) = .3$, what conclusion can you make about A and B?
   (1) $P(B|A) = .6$
   (2) A and B are independent events.
   (3) A and B are not independent events.
   (4) No conclusion is possible.

7. If $P(A) = .7$, $P(B) = .6$, and $P(A$ and $B) = .42$, what conclusion can you make?
   (1) $P(A|B) = .6$
   (2) A and B are independent events.
   (3) A and B are not independent events.
   (4) No conclusion can be made.

8. A group of 40 adults are asked if they like classical music, rock music, both, or neither. The results are collected in this Venn diagram.

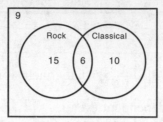

Based on this data, what conclusion can be made about liking classical music A and liking rock music B?

(1) $P(B) \neq P(B|A)$

(2) They are not independent of each other.

(3) They are independent of each other.

(4) No conclusion can be made.

9. 60 students are surveyed and asked if they use Facebook A, Instagram B, both, or neither. The results of the survey are collected in this Venn diagram below.

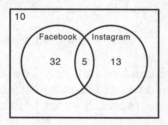

Based on this data, what conclusion can be made about using Facebook and using Instagram?

(1) $P(A|B) = P(A)$

(2) They are not independent of each other.

(3) They are independent of each other.

(4) No conclusion can be made.

10. 200 people are surveyed and asked two questions. (1) Which drink do you prefer: water, juice, or milk? (2) Which snack do you prefer: fruit or chips? The data are shown on the table below.

	Water	Juice	Milk	Total
Fruit	50	70	10	130
Chips	20	10	40	70
Total	70	80	50	200

Based on the data in the table, which of the following is *not* true?

(1) $P(\text{prefers juice or prefers chips}) = \dfrac{150}{200}$

(2) $P(\text{prefers fruit and prefers water}) = \dfrac{50}{200}$

(3) $P(\text{prefers fruit given that they prefer water}) = \dfrac{50}{70}$

(4) $P(\text{prefers milk given that they prefer chips}) = \dfrac{40}{70}$

B. *Show how you arrived at your answers.*

1. A is the event "Audrina smokes a pack of cigarettes a day." What is an event that is likely *not* to be independent of A?

2. $P(A) = .2$, $P(B) = .7$. How can the value of $P(A \text{ and } B)$ help you determine if A and B are independent events?

3. The probability that Charlotte goes to the party is P(Charlotte goes to party) = .7. The probability that Layla goes to the party is P(Layla goes to party) = .6. Explain how it could be that P(Layla goes to the party given that Charlotte goes to the party) = .9?

4. 90 students were polled about whether they like soccer, basketball, both, or neither. Based on this data, determine if liking basketball and liking soccer are independent events.

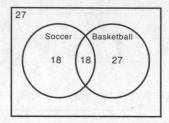

5. 150 people are surveyed and asked these two questions. (1) Which subject do you prefer: math, English, or science? (2) Is your hair curly or straight? The data are shown on the chart below.

	Math	**English**	**Science**	**Total**
Curly hair	25	30	35	90
Straight hair	45	5	10	60
Total	70	35	45	150

Based on the data in the chart, is having straight hair independent of preferring science? Explain your reasoning.

NORMAL DISTRIBUTION

Chapter Ten

10.1 STANDARD DEVIATION

∧
KEY IDEAS
∠——∖

The *standard deviation* of a set of numbers is a measure of how "spread out" the numbers in the set are. If all the numbers are equal, the standard deviation is zero. The more spread out the numbers are, the larger the standard deviation.

Comparing Standard Deviations for Two Data Sets with the Same Mean

Two rooms each have 100 people in them. The mean (average) height of the people in each room is approximately 66 inches. Even though the rooms have the same number of people and the same mean, there is something very different about the collection of people in each room.

Below are the dot plots for room 1 and room 2.

Room 1

Height

Room 2

Height

The people in room 1 are mostly very close to the 66-inch average. There are people in room 2, though, who are much shorter than the average and people who are much taller than the average.

In statistics, we say that the set of people in the second room have a higher standard deviation. The standard deviation for the heights of the people in room 1 was approximately 2, while the standard deviation for the heights of the people in room 2 was 6.

Example 1

Each of the dot plots below has a mean of approximately 40. Which of the dot plots below corresponds to the set with the lowest standard deviation?

(1)

(2)

(3)

(4)

Solution: Choice (4) is correct. The dots are bunched together much more closely in choice (4) than in the other choices. Choice (1) has a standard deviation of about 7. Choice (2) has a standard deviation of about 5. Choice (3) has a standard deviation of about 3. Choice (4) has a standard deviation of about 1.

Calculating Standard Deviation with a Calculator

On 10 Algebra II quizzes, Willian gets scores of 72, 76, 81, 81, 79, 71, 85, 79, 85, and 89. Calculating the standard deviation for the quiz scores can be done very quickly with a graphing calculator.

For the TI-84:

Press [STAT] [1] to get to the list editor. Go to the L1 cell, and press [CLEAR] [ENTER]. Enter the 10 scores into L1. Press [STAT], move to "CALC," and select [1]. The first value, 81.1 is the mean. The fifth value, next to the σx=, is the standard deviation. The standard deviation for this data set is 5.4.

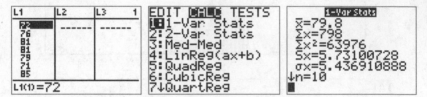

For the TI-Nspire:

From the home screen, select the Add Lists & Spreadsheet icon. Name column A *x*, and fill in cells A1 though A10 with the 10 scores. Press [menu] [4] [1] [1] for One-Variable Statistics. Select the [ok] button. Set the X1 List to a[ ]. Set Frequency List to 1. In cell E2 will be the mean, 81.1. In cell C6 will be the standard deviation next to the σx=. The standard deviation for this data set is 5.4.

Example 2

Find the standard deviation of the data set 4, 6, 8, 8, 9, 12, 15, rounded to the nearest tenth.

Solution: Using the graphing calculator, the standard deviation of this set is 3.4.

Standard Deviation for Sets with Repeated Numbers

Entering all the numbers for a large data set into the calculator can be very time consuming. If, however, there are many repeated numbers, you can use a shortcut. Below the data for 55 numbers are summarized in a *frequency table*. The 7 in the frequency column after the number 42 means that there are seven 42s in the data set.

Number	Frequency
42	7
53	9
61	12
74	8
86	4
127	10
150	5

Though this could be done by entering 42, 42, 42, 42, 42, 42, 42, 53, 53, 53, etc., into the calculator, there is a shortcut for finding the standard deviation of the complete 55-number set.

For the TI-84:

Press [STAT] [1] and enter the numbers 42, 53, 61, 74, 86, 127, and 150 into L1, and enter the numbers 7, 9, 12, 8, 4, 10, 5 into L2. Press [STAT], move to "CALC" and press [1]. In the next menu, put L1 into the List field by pressing [2ND][1] and L2 into the FreqList field by pressing [2ND][2]. Then select Calculate. The standard deviation is 35.2.

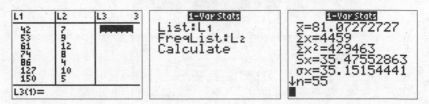

For the TI-Nspire:

From the home screen, select the Add Lists & Spreadsheet icon. Name column A x, and fill in cells A1 though A7 with the 7 scores. Name column B y, and fill in cells B1 through B7 with the 7 frequencies. Press [menu] [4] [1] [1] for One-Variable Statistics. Set the number of lists to 1. Select the [ok] button. Set the X1 List to a[ ] and the Frequency List to b[ ]. In cell E2 will be the mean, 81.1. In cell E6 will be the standard deviation next to the $\sigma x=$. The standard deviation for this data set is 35.2.

z-scores

In a data set with a mean of 60 and a standard deviation of 4, the number 64 is one standard deviation above the mean since it is equal to the mean plus the standard deviation, $60 + 4 = 64$. The number 68 is two standard deviations above the mean since it is equal to the mean plus two times the standard deviation $60 + 2 \cdot 4 = 68$.

In statistics, we say that a number that is one standard deviation above the mean has a *z-score* of 1, while a number that is two standard deviations above the mean has a *z*-score of 2.

MATH FACTS

The *z-score* of a number is how many standard deviations above or below the mean that number is. If the number is greater than the mean, the *z*-score is positive. If the number is less than the mean, the *z*-score is negative.

A formula for calculating *z*-score is $z = \dfrac{\text{number} - \text{mean}}{\text{standard deviation}}$.

Example 3

If the mean is 60 and the standard deviation is 4, what is the *z*-score of the number 52?

Solution: Since 52 is less than 60, the *z*-score will be negative. For this example, it is possible to subtract the standard deviation, 4, from 60 to get to 56, which has a *z*-score of -1. Then subtract it again to get to 52, which has a *z*-score of -2.

This can also be calculated with the formula $z = \dfrac{\text{number} - \text{mean}}{\text{standard deviation}} = \dfrac{52 - 60}{4} = \dfrac{-8}{4} = -2$.

Example 4

If the mean is 40 and the standard deviation is 7, what is the *z*-score of the number 50?

Solution: The *z*-score must be between 1 and 2 since $40 + 7 = 47$ would have a *z*-score of 1 while $40 + 2 \cdot 7 = 54$ would have a *z*-score of 2.

Use the formula: $z = \dfrac{50 - 40}{7} = \dfrac{10}{7} \approx 1.4$

Example 5

If the mean is 73 and the standard deviation is 8, what number has a z-score of −1.5?

Solution: Multiply the z-score by the standard deviation to get −1.5 · 8 = −12. Add this to the mean to get 73 + (−12) = 61, which is the number that has a z-score of −1.5.

This can also be done with the formula with z as a known variable and the number as the unknown.

$$-1.5 = \frac{x - 73}{8}$$

$$8 \cdot (-1.5) = x - 73$$

$$-12 = x - 73$$

$$+73 = +73$$

$$\overline{61 = x}$$

Check Your Understanding of Section 10.1

A. *Multiple-Choice*

1. Each of these data sets has a mean of 5. Which has the smallest standard deviation?
   (1) 1, 2, 5, 7, 10   (3) 4, 5, 5, 5, 6
   (2) 2, 3, 5, 5, 10   (4) 3, 4, 5, 5, 8

2. What is the standard deviation of the data set 16, 19, 14, 12, 10, 15, 15, 16?
   (1) 2.5   (2) 2.6   (3) 2.7   (4) 2.8

3. What is the standard deviation for this data set?

x_i	f_i
18	4
19	7
20	2
21	3
22	6

   (1) 1.41   (2) 1.51   (3) 1.54   (4) 1.61

4. If the mean of a data set is 60 and the standard deviation is 8, what is the z-score for the number 72?
   (1) .67   (2) 1.00   (3) 1.25   (4) 1.50

5. If the mean of a data set is 40 and the standard deviation is 6, what number has a z-score of −2.5?
   (1) 25   (2) 35   (3) 45   (4) 55

6. Which number has a z-score closest to −1 in the data set 45, 38, 47, 41, 42, 34, 46, 37?
   (1) 42   (2) 41   (3) 38   (4) 37

7. If the mean of a data set is 41 and the standard deviation is 6, which of the following has a z-score between 1.5 and 2.0?
   (1) 45   (2) 47   (3) 49   (4) 51

8. How many of these 8 numbers are more than two standard deviations from the mean?

   $$18, 40, 44, 45, 48, 50, 50, 68$$

   (1) 1   (2) 2   (3) 3   (4) 0

9. How many of these 8 numbers are more than two standard deviations from the mean?

   $$11, 49, 52, 52, 53, 53, 55, 90$$

   (1) 1   (2) 2   (3) 3   (4) 0

10. What percent of these 14 numbers have a z-score between −1 and +1?

   $$59, 62, 54, 58, 62, 62, 55, 56, 65, 64, 58, 53, 59, 61$$

   (1) 68%   (2) 71%   (3) 78%   (4) 93%

10.2 NORMAL DISTRIBUTION

∧
KEY IDEAS
∠   ∖

The dot plot of real-world data often has the shape of a *bell curve*, which is also known as a *normal distribution curve*. When this happens, it is possible to answer certain questions about the data that you would not be able to had the shape of the dot plot not been a bell curve.

The Normal Curve

Below is a dot plot with 400 data points representing the weights of 400 children. The mean value is approximately 60, and the standard deviation is approximately 4.

Height

Since the mean is 60 and the standard deviation is 4, a score of 64 is said to be *one standard deviation above the mean*. A score of 68 is said to be *two standard deviations above the mean*. A score of 56 is said to be *one standard deviation below the mean*. A score of 52 is said to be *two standard deviations below the mean*.

Percentages in the Normal Distribution Curve

In a normal distribution with a lot of data points, approximately 68% of the values will be between one standard deviation below the mean and one standard deviation above the mean. 95% of the values will be between two standard deviations below the mean and two standard deviations above the mean.

310

The dot plot about the weights of the 100 children was approximately *normal* because 70% of the 100 points are between 56 and 64 while 93% of the 100 points are between 52 and 68.

Using a Calculator to Find the Percentage of Numbers Between Two Values

If the mean and the standard deviation of a data set are known, the graphing calculator can quickly determine what percent of the numbers in the data set will be between any two values.

The heights of 400 people are measured. The mean is approximately 60, and the standard deviation is approximately 4.

Height

Since 68% of the numbers are expected to fall within one standard deviation of the mean, then approximately $.68 \cdot 400 = 272$ should be between $60 - 4 = 56$ inches and $60 + 4 = 64$ inches. Since 95% of the numbers are expected to fall within two standard deviations of the mean, then approximately $.95 \cdot 400 = 380$ should be between $60 - 8 = 52$ inches and $60 + 8 = 68$ inches.

Knowing the 68% and 95% rule is useful if a question asks how many members of a set are within one or two standard deviations of the mean. Most questions, though, are about finding how many members of a set are between two numbers that are not exactly one or two standard deviations from the mean.

In the scenario above, a graphing calculator can be used to determine how many out of the 400 people can be expected to be between 57.5 and 62 inches tall. Between 56 and 64 inches was 68% since those were the numbers exactly one standard deviation above and below the mean. The percent of people between 57.5 and 62 inches will be lower than 68% and can be found with the graphing calculator.

For the TI-84:

Press [2ND] [VARS] [2] for normacdf. Fill in the fields with 57.5 for lower, 62 for upper, 60 for μ, and 4 for σ Then select "PASTE."

```
DISTR DRAW          normalcdf         normalcdf(57.5,▶
1:normalpdf(        lower:57.5                .425477
2:normalcdf(        upper:62
3:invNorm(          μ:60
4:invT(             σ:4
5:tpdf(             Paste
6:tcdf(
7↓χ²pdf(
```

For the TI-Nspire:

From the home screen, press [A]. Press [menu] [6] [5] [2] for Normal Cdf. Set the Lower Bound to 57.5, the Upper Bound to 62, the mean (μ) to 60, and the standard deviation (σ) to 4.

```
1: Actions
1: Normal Pdf...                    DEG
2: Normal Cdf..
3: Inverse Normal...
4: t Pdf...
5: t Cdf...          1: Stat Calculations
6: Inverse t...      2: Stat Results
7: χ² Pdf...         3: List Math
8: χ² Cdf...         4: List Operations
9: Inverse χ²...     5: Distributions
A: F Pdf...          6: Confidence Intervals
                     7: Stat Tests
```

```
Scratchpad ▽            DEG
Normal Cdf

Lower Bound:  57.5
Upper Bound:  62
         μ    60
         σ    4

                  OK  Cancel
```

```
Scratchpad ▽            DEG

normCdf(57.5,62,60,4)       0.425477
```

The percent is approximately 43%, which is approximately .43 · 400 = 72 people between 57.5 inches and 62 inches tall.

Example 1

If the mean of a set of 500 Regents scores is 80, the standard deviation is 15, and the data are approximately normally distributed, approximately how many scores can be expected to be between 74 and 89?

Solution: Using the Normal cdf function of the calculator, the percent of scores between 74 and 89 will be approximately 38%. Out of 500 scores, this will be approximately .38 · 500 = 190 scores.

Example 2

If the mean temperature in New York throughout the year is 55 degrees and the standard deviation is 15 degrees, what percent of the days can be expected to have temperatures greater than 80 degrees?

Solution: In this example, there is no upper bound. When this happens, make the upper bound a very large number, at least four standard deviations above the mean. (The number 9,999 is generally safe to use unless there is a very large mean and/or standard deviation.) When using 9,999 as the upper bound, the calculator determines the percent to be approximately 5%.

For examples where there is no lower bound, make the lower bound a very small negative number like −9,999.

Check Your Understanding of Section 10.2

A. Multiple-Choice

1. If the mean of a set of normally distributed 200 numbers is 70 and the standard deviation is 6, approximately what percent of the numbers will be between 64 and 76?
   (1) 68%   (2) 70%   (3) 76%   (4) 80%

2. The weights of 400 children are collected, and the data are normally distributed. What percent of the children were between 42 and 52 pounds if the mean was 50 pounds and the standard deviation was 3?
   (1) 57%   (2) 60%   (3) 63%   (4) 74%

3. On the Algebra II Regents, the mean score was 75 and the standard deviation was 9. If the data are normally distributed, what percent of the scores were over 90?
   (1) 20%   (2) 15%   (3) 10%   (4) 5%

4. The amount of time that 300 adults exercise per week was collected. If the data were normally distributed, what percent of the people exercised under 20 minutes a week if the mean was 40 minutes and the standard deviation was 15?
   (1) 9%   (2) 11%   (3) 13%   (4) 15%

5. In a normally distributed data set, what percent of the samples will have a *z*-score between 1 and 2?
   (1) 14%   (2) 16%   (3) 18%   (4) 20%

6. In a set of 500 samples, the mean is 90 and the standard deviation is 17. If the data are normally distributed, how many of the 500 are expected to have a value between 93 and 101?
   (1) 82   (2) 84   (3) 86   (4) 88

7. In a high school, 1,000 students run a mile. The mean time is 9 minutes with a standard deviation of 2 minutes. What running time would put a student in the top 5 percentile rank?
   (1) 4.3 minutes   (3) 5.3 minutes
   (2) 4.8 minutes   (4) 5.8 minutes

8. A group of baseball players had a mean of 15.3 home runs with a standard deviation of 9.9. If the data are normally distributed, what percent of players can be expected to hit over 30 home runs?
   (1) 3%   (2) 5%   (3) 7%   (4) 9%

9. The average snowfall in Regents Town is 19 inches per year with a standard deviation of 5 inches. If the data are normally distributed, what is the probability that it will snow between 18.5 and 19.5 inches in a year?
   (1) 6%   (2) 8%   (3) 20%   (4) 27%

10. In a normally distributed data set with a mean of 60, what is more likely: a number greater than 80 or a number less than 30?
    (1) A number greater than 80
    (2) A number less than 30
    (3) They are equally likely
    (4) It depends on what the standard deviation is

B. Show how you arrived at your answers.

1. A data set of ten numbers is shown below:

   28, 83, 73, 88, 81, 54, 75, 66, 66, 91

   a) Calculate the mean and the standard deviation for the data set.
   b) How many of the numbers fall within one standard deviation of the mean?
   c) Based on your answer to part (b), do you think this data set is approximately normal or not? Explain your reasoning.

2. Which, if any, of these numbers is not within 2 standard deviations of the mean?

$$46, 47, 38, 38, 36, 41, 44, 40, 33, 38, 40, 39?$$

3. The heights of 200 adults are measured. The average height is 66 inches, and the standard deviation is 8 inches. If 14 of the people are over 82 inches, what does this suggest about whether or not the data are normally distributed? Explain your reasoning.

4. The ages of 20 people are shown:

$$66, 72, 70, 74, 78$$

$$80, 73, 68, 84, 76$$

$$76, 78, 70, 62, 75$$

$$76, 79, 73, 73, 70$$

Which, if any, of these ages are not within 2 standard deviations of the mean?

5. A data set has 68.3% of the population of samples within one standard deviation of the mean. Is this a normal distribution? Explain your reasoning.

STATISTICS

11.1 TYPES OF STATISTICAL STUDIES

$$\stackrel{\wedge}{\underline{\underline{\qquad\text{KEY IDEAS}\qquad}}}$$

Three ways to collect data for a statistical study are *survey*, *observational study*, and *experiment*. *Bias* in a statistical study is when something about how the data are collected may have caused inaccurate results. Causes of this bias should be identified and, if possible, eliminated before collecting the data.

Statistical Survey

A survey is a question or series of questions that participants, also called *subjects*, in the study are asked to answer. The survey can have a simple yes/no question like "Do you like pistachio ice cream?" or a question with a numerical answer like "How much time did you spend on homework last night?" A survey is the simplest way of collecting data, but there are many ways that it can lead to bias.

The wording of the survey question can cause bias. For example, if the question says "Do you not hate pistachio ice cream?" rather than "Do you like pistachio ice cream?" the negative phrasing of the question could change the results even if the questions are supposed to mean the same thing.

The participants of the survey should be randomly selected, otherwise the results can also be distorted. The results of a survey about pop music, for example, will not be accurate if most of the people asked to complete the survey are under 20 years old. If the participants in the survey do not accurately represent the total population, the survey has *selection bias*.

If the survey is voluntary, there is a chance that people who respond to the survey are more likely to answer the questions a certain way. A survey that is conducted by text messaging, for example, might be answered more by younger people, causing bias in the results.

Observational Study

An *observational study* is like a survey. Instead of subjects being asked to answer questions, the person conducting the study observes the behavior of the participants and records the results. To learn about whether or not people like pistachio ice cream, the observer could go to an ice cream shop and watch what different people order. An important aspect of an observational study is that the person conducting the study cannot do anything that could interfere with or control what the subjects do.

An observational study could have bias if the subjects are not randomly selected. For example, if the ice cream shop observations happen during school hours, people under 18 will not be adequately represented.

Experimental Study

In an *experimental study*, the person conducting the study randomly chooses some of the subjects and exposes them to some kind of *treatment*. An example is a study to see if taking vitamin C prevents colds. Fifty people are randomly selected and then, from those fifty, twenty-five are randomly selected to take vitamin C pills daily while the other twenty-five are not given the pills.

It is important in an experimental study to choose the group of people getting the treatment randomly. If there is some kind of bias in who receives the treatment, the experiment might lead to an inaccurate conclusion.

Check Your Understanding of Section 11.1

B. Show how you arrived at your answers.

1. You want to find out what type of movie is the favorite among the residents of New York City. If you want to do this by conducting a survey, what are some ways that you can reduce bias?

2. You want to study how the amount of sleep a student gets the night before the Regents relates to his or her Regents score. How can this be studied as an observational study? How can this be studied as an experiment?

3. There are 500 10th graders at Regentsville High School. You want to do an experiment where 20 students study for the Regents with just

the red Barron's book and 20 other students study for the Regents with both the blue and the red Barron's books. How can the groups be chosen to reduce the chance of skewed data through bias?

4. A survey is conducted to learn what subjects students enjoy most at school. One question reads, "What is your most boring subject?" Could the wording of this question skew the results of the survey? Explain.

5. Is this describing a survey, an experiment, or an observational study? Explain.

A teacher teaches two Algebra II math classes, one during period 5 and another during period 6. In the period 5 class, the temperature in the room is set to 70 degrees for a two-week period. In the period 6 class, the temperature in the room is set to 85 degrees for the same two-week period. At the end of the two weeks, the students are given a test to see how well they learned the math. The results are collected and analyzed.

11.2 INFERENTIAL STATISTICS

⋀ KEY IDEAS ⋀

If much of the data in a data set is unknown, it is still possible to estimate different information about it by using a smaller collection of the total set known as a *sample*. Statistics about the sample can provide valuable information about the complete data set.

A complete data set can contain thousands, if not millions, of data points. The complete set of data is known as the *population*. A much smaller subset of the population is known as a *sample*.

Two types of measures about the population that can be estimated with samples are the *population proportion* and the *population mean*. The population proportion is the percent of the total number of things in the population that has a certain characteristic. The population mean is the average value of some characteristic of all the things in the population. The percent of some sample of things that has a certain characteristic is called the *sample proportion*. The *sample mean* is the average value of some characteristic of all the things in that sample.

To estimate the population proportion or the population mean, you generally find a sample proportion or sample mean and then add and subtract the *standard error*.

For example, if the sample mean is 20 and the standard error is 7, the population mean will be between $20 - 7 = 13$ and $20 + 7 = 27$. If the sample proportion is 0.40 and the standard error is 0.15, the population mean estimate will be between $0.40 - 0.15 = 0.25$ and $0.40 + 0.15 = 0.55$.

Calculating the standard error depends on the type of question. Four different types are presented below.

Using One Sample Set to Approximate the Population Proportion

In an auditorium with 800 students, a certain percent of them will be boys and a certain percent will be girls. If 450 of those students are boys, then the percent of total students who are boys is $\frac{450}{800} = 0.5625 = 56.25\%$.

This percent is the population proportion, represented by the symbol p, of boys for the entire population of students.

If, for whatever reason, we are only able to survey a subset of 75 students, we can calculate the sample proportion for that sample. If 45 out of those 75 students were boys, the sample proportion for the percent of boys, represented by the symbol $\hat{p}$, would be $\hat{p} = \frac{45}{75} = 0.60$.

For this type of problem, the standard error, SE, can be calculated with the following formula.

$$SE = 2 \cdot \sqrt{\frac{\hat{p}(1-\hat{p})}{n}} = 2 \cdot \sqrt{\frac{0.60(1-0.60)}{75}} \approx 0.11$$

The sample proportion is different from the population proportion in this example. However, there is a formula that allows you to find a range of likely values for the population proportion based on this one sample of 75 students.

$$\hat{p} - SE \le p \le \hat{p} + SE$$

The population proportion is, *almost surely*, between $0.60 + 0.11 = 0.71 = 71\%$ and $0.60 - 0.11 = 0.49 = 49\%$.

This is a pretty large range of values. However, the actual value of the population proportion, 56.25%, is between 49% and 71% as the formula predicted.

Using Multiple Sample Sets to Approximate the Population Proportion

For a more accurate estimate of the population proportion, more than one sample set is needed. In the auditorium example, one sample set of 75 students with a sample mean of 0.60 was used. To get a better approximation, repeat this process 50 times. The first sample of 75 had 45 boys, which was 60%. The second sample of 75 had 35 boys, which was approximately 47%. A computer is used to do this process 50 times. When the percent of boys for all 50 sets of 75 students are plotted on a dot plot, it looks like this.

Measures from Sample of Collection 1

proportion of boys

This plot is called the *sampling distribution of the sample proportion*.

Two statistics about this plot are also provided. The mean value for the numbers in this plot is approximately 0.54, and the standard deviation for the numbers in this plot is approximately 0.05.

These two numbers can now be used to approximate the population proportion of the complete data set.

For this type of problem, the standard error, SE, can be calculated with the following formula.

$$SE = 2 \cdot \sigma = 2 \cdot 0.05 = 0.10$$

The estimated population proportion is then mean of plot − SE ≤ p ≤ mean of plot + SE.

For this example the actual value of the population proportion is, *almost surely*, between 0.54 − 0.10 = 0.44 and 0.54 + 0.10 = 0.64. The actual proportion is 56.25%, which is, indeed, between 44% and 64%.

On the Regents, the information will be provided so you can do the analysis. Creating the dot plot and finding the mean and the standard deviation of the numbers in the dot plot is something that requires a computer.

Making this plot required 50 · 75 = 3,750 numbers. Even though this is a lot of numbers for calculating the percent of boys out of just 800 people, this method of using a computer to do this analysis is very powerful either when the population is either extremely large or when, for various reasons, it is not possible to get access to all the data.

Using One Sample Set to Approximate the Population Mean

At a high school graduation party are 600 people, including parents and students. There are more parents than students at the party, and a dot plot of the ages of all the party goers looks like this.

Graduation Party

age

The approximate average age of the 600 people at the party can be approximated by first taking a random sample of 30 people at the party. Here is one random sampling of 30 people from the total population of 600 people.

51, 45, 42, 13, 40, 20, 45, 49, 41, 15, 18, 45, 41, 14, 35, 41, 17, 40, 42, 45, 51, 18, 31, 11, 19, 47, 20, 14, 17, 47

To use these 30 numbers to approximate the mean age of the 600 people, first find the mean and the standard deviation of these 30 values. For this sample set, the mean value, represented by $\bar{x}$, is approximately 32.4. The standard deviation of the sample, represented by s, is approximately 13.8.

For this type of problem the standard error, SE, can be calculated with the following formula.

$$\text{SE} = 2 \cdot \frac{s}{\sqrt{n}} = 2 \cdot \frac{13.8}{\sqrt{30}} \approx 5.04$$

The formula for approximating the population mean, represented by μ, is $\bar{x} - \text{SE} \leq \mu \leq \bar{x} + \text{SE}$.

So according to this formula, the population mean is *almost surely* between $32.4 - 5.04 = 27.4$ years old and $32.4 + 5.05 = 37.4$ years old. The actual population mean is 34.4 years old, so the predicted range was correct.

Using Multiple Samples Set to Approximate the Population Mean

By using just one sample set of 30 ages, it was possible to find an approximate value of the population mean. To get an even more accurate approximation, more sample sets are needed.

By using a computer, 100 sample sets of 30 ages each were taken. The mean for the first set, used in the other approximation, was 32.4. The mean for the second set was 36.5. The mean for each of these 100 sets was then plotted on this dot plot or histogram.

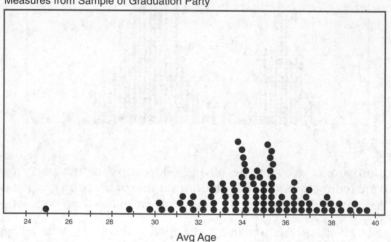

Measures from Sample of Graduation Party

Avg Age

Measures from Sample of Graduation Party

This is called the *sampling distribution of the sample mean.*

Two more statistics will be needed and likely given on the Regents. The mean value for the numbers in the plot is approximately 34.5. The standard deviation is approximately 2.5.

For this scenario, the standard error, SE, is 2 times the standard deviation of all the different means.

$$SE = 2 \cdot 2.5 = 5$$

With these numbers, the population mean can be calculated with the formula:

$$\text{mean of plot} - SE \le \mu \le \text{mean of plot} + SE.$$

For this example, $34.5 - 5 \le \mu \le 34.5 + 5$. So the population mean is between 29.5 and 39.5. This is more accurate than the approximation that used just one set of 30. The mean of the samples, 34.5, is extremely close to the actual population mean of 34.4.

If the standard deviation of the sample means is not given, it can be approximated with the formula $\sigma = \dfrac{s}{\sqrt{n}}$, where s is the standard deviation of any one of the sample sets and n is the number of samples in each of the sample sets. When using the first sample set, which had a standard deviation of 13.8, $\sigma = \dfrac{13.8}{\sqrt{30}} \approx 2.5$. Since the standard error is two times the standard deviation, it is also true that $SE = 2 \cdot \dfrac{s}{\sqrt{n}}$.

Using a Simulation for a Randomization Test

A *randomization test* is a way to check if the results of a small random sample are near what you would expect (or hope) them to be.

In this example, a focus group of 30 students are asked to give their opinion on a new style of backpack. The manufacturer is willing to mass

323

produce the backpack if 40% of students like it. Unfortunately, only 9 of the 30 students polled, or 30%, liked the new backpack.

There is a way to check if that 30% is close enough to the 40% of the entire student population the manufacturer had specified. A *randomization test* can be used. This is similar to the process for estimating the population mean from a bunch of random sample sets.

A computer is used to generate 100 random sets of 30 students from an imaginary group of students from which exactly 40% of them do like the backpack.

The first set of these imaginary 30 students happened to have 37% liking the backpack. This process is continued 100 times. The percent who like the backpack for each sample set is graphed on a dot plot or histogram. The mean and the standard deviation of those numbers are calculated.

Measures from Sample of Collection

proportion of ones

In this case, the mean value is 0.404 and the standard deviation is 0.089.

Even without being told the mean and standard deviation of this plot of the sampling distribution for the imaginary set, it is possible to approximate them. The mean will be 0.40, since that is the proportion that was used to create the plot. The standard deviation can be calculated with the formula $\sigma = \dfrac{\sqrt{p(1-p)}}{n} = \dfrac{\sqrt{0.4 \cdot 0.6}}{30} \approx 0.089$, which was exactly what it was for the data on the dot plot. On the dot plot, it can be seen that 30% or less happened 16 times. So it does not seem so unusual for this to happen.

To determine if 30% is something that is *likely* to happen even though the real percent was 40% for this imaginary population, check to see if the 30% is within two standard deviations of the mean.

For this example, $0.404 - 2 \cdot 0.089 = 0.226$ and $0.404 + 2 \cdot 0.089 = 0.582$. Since 30% is between 22.6% and 58.2%, the company can justify producing the backpacks even though the sample of 30 people had fewer than 40% liking it.

Check Your Understanding of Section 11.2

A. Multiple-Choice

1. An ecologist wants to know what percent of the 10,000 fish in a lake are cod. She conducts a study in which 50 fish are randomly caught. Of those 50 fish, 15 are cod. Based on this data, using a 95% confidence interval, the researcher can determine that the percent of fish that are cod in the lake is which of the following?
   (1) Between 17% and 43%   (3) Between 24% and 36%
   (2) Exactly 30%   (4) Between 20% and 40%

2. Out of 30,000 households in a town, 100 are selected at random and asked if they have a landline or not. 70 out of the 100 households chosen replied that they do have a landline. Based on this data, using a 95% confidence interval, you can determine that the percent of households in the town that have a landline is which of the following?
   (1) Exactly 70%   (3) Between 64% and 76%
   (2) Between 68% and 72%   (4) Between 61% and 79%

3. A sports physician conducts an observational study to learn the average amount of time that 3,000 swimmers in the town can hold their breath underwater. 60 swimmers are chosen at random and tested. The average amount of time those 60 swimmers can hold their breath is 70 seconds with a standard deviation of 6. Based on this data, using a 95% confidence interval, the researcher can determine that for the entire 3,000 swimmers, the average amount of time they can hold their breath underwater is which of the following?
   (1) Between 71.5 seconds and 68.5 seconds
   (2) Exactly 70 seconds
   (3) Between 69.5 seconds and 70.5 seconds
   (4) Between 69 seconds and 71 seconds

4. A concert promoter wants to estimate the average age of the 20,000 people attending a Taylor Swift concert. At the concert, 80 people are selected at random. The average age of the 80 people is 24.3 with a standard deviation of 11.9. Based on this data, using a 95% confidence interval, the researcher can determine that the average age for all 20,000 people at the concert is which of the following?
(1) Exactly 24.3
(2) Between 21.64 and 26.96
(3) Between 23.3 and 25.3
(4) Between 23.8 and 25.1

5. Use the same population as from question 1 except take 500 samples of size 50. The average for all these samplings is 0.24 with a standard deviation of 0.06.

This is a histogram of the sampling distribution of the sample proportion.

Using this data, with a 95% confidence interval, we can determine that the percent of fish in the lake that are cod is which of the following?
(1) Between 0.12 and 0.36
(2) Between 0.14 and 0.34
(3) Between 0.20 and 0.28
(4) Exactly 0.24

6. Using the same population as in question 2 but take 300 samplings of size 100. The average of all the means is 0.73, and the standard deviation is 0.05.

This is a histogram of the sampling distribution of the sample proportion.

Using this data, with a 95% confidence interval, we can determine that the percent of homes in the town that have a land line is which of the following?
(1) Between 0.63 and 0.83   (3) Between 0.70 and 0.76
(2) Between 0.68 and 0.78   (4) Exactly 0.73

7. The sports physician uses the same population as in question 3 but uses 150 samplings of 60 people. The average of the means of all the samplings is 72.7, and the standard deviation is 0.92.

This is a histogram of the sampling distribution of the sample mean.

Based on this data, with a 95% confidence interval, the researchers can determine that the actual average amount of time the entire population can hold its breath under water is which of the following?

(1) Exactly 72.7

(3) Between 71.28 and 73.12

(2) Between 72 and 73.4

(4) Between 70.86 and 74.54

8. The concert promoter uses the same population as in question 4 but takes 150 samplings of 80 people. The average of the means of all the samplings is 25.5, and the standard deviation is 1.5.

This is a histogram of the sampling distribution of the sample mean.

Based on this data, with a 95% confidence interval, the researchers can determine that the average age of the entire 20,000 person population is which of the following?

(1) Exactly 25.5

(3) Between 23.5 and 27.5

(2) Between 22.5 and 28.5

(4) Between 24.5 and 26.5

9. Amy Hogan rolls a six-sided die 12 times and lands on 6 on four of the rolls. She wants to examine if this might suggest that the die is defective since, on average, the number of expected 6s in 12 rolls is 2. She uses a computer to simulate 500 times of rolling a fair die 12 times. A histogram of the results is shown below.

Based on this histogram, Amy concludes which of the following?
(1) Yes, the die is defective because two 6s is most common.
(2) Yes, the die is defective because four 6s is not common.
(3) No, the die is not necessarily defective because two 6s is not very common.
(4) No, the die is not necessarily defective because four 6s is fairly common.

10. A new plant food is developed that the inventors claim helps plants grow faster. They cite an experiment they did where out of 40 randomly chosen plants, 20 of them were randomly chosen to receive the new plant food while the other 20 plants received a different kind of plant food.

The 20 plants that received the new plant food grew an average of 2 inches more than the plants that received a different plant food. To test if this difference is statistically significant, the data for the 40 plants are entered into a computer that randomly separates them into two groups of 20. This is done 50 times. The results are shown in the histogram below, where the number is the difference between the first group and the second group of the randomly chosen groupings.

Based on this data, can you conclude that the new plant food causes plants to grow faster?
(1) No, the difference is not statistically significant. +2 is uncommon.
(2) No, the difference is not statistically significant. +2 is fairly common.
(3) Yes, the difference is statistically significant. +2 is uncommon.
(4) Yes, the difference is statistically significant. +2 is fairly common.

TEST-TAKING STRATEGIES

Knowing the material is only part of the battle in acing the Algebra Regents exam. Things like not managing your time, making careless errors, and struggling with the calculator can cost valuable points. In this chapter some test-taking strategies are explained to help you perform your best on test day.

12.1   TIME MANAGEMENT

Don't Rush

The Algebra Regents exam is three hours long. Even though you are permitted to leave after one and a half hours, to get the best grade possible, stay until the end of the exam. Just as it wouldn't be wise to come to the test an hour late, it is almost the same thing if you leave a test an hour early.

Do the Test Twice

The best way to protect against careless errors is to do the entire test twice and compare the answers you got the first time to the answers you got the second time. For any answers that don't agree, do a "tie breaker" third time. Redoing the test and comparing answers is much more effective than simply "looking over" your work. People tend to skim by careless errors when looking over their work. Redoing the questions, you are less likely to make the same careless error.

Bring a Watch

In some classrooms the clock is broken. Without knowing how much time is left, you might rush and make careless errors. Though the proctor should write the time on the board and update it every so often, it is best if you have your own watch.

The TI-84 graphing calculator has a built in clock. Press the [MODE] to see it. If the time is not right, go to SET CLOCK and set it correctly. The TI-Nspire does not have a built-in clock.

```
NORMAL  SCI  ENG          FORMAT:M/D/Y D/M/Y Y/M/D
FLOAT 0123456789          YEAR:  2014
RADIAN DEGREE             MONTH: 7
FUNC  PAR  POL  SEQ       DAY:   28
CONNECTED  DOT            TIME:  12HOUR 24HOUR
SEQUENTIAL  SIMUL         HOUR:  9
REAL  a+bi  re^θi         MINUTE: 32
FULL  HORIZ  G-T          AM/PM:  AM PM
SET CLOCK 07/28/14 9:32AM
```

12.2   KNOW HOW TO GET PARTIAL CREDIT

Know the Structure of the Exam

The Algebra Regents exam has 37 questions. The first 24 of those questions are multiple-choice worth two points each. There is no partial credit if you make a mistake on one of those questions. Even the smallest careless error, like missing a negative sign, will result in no credit for that question.

Parts Two, Three, and Four are free-response questions with no multiple-choice. Besides giving a numerical answer, you may be asked to explain your reasoning.

Part Two has eight free-response questions worth two points each. The smallest careless error will cause you to lose one point, which is half the value of the question.

Part Three has four free-response questions worth four points each. These questions generally have multiple parts.

Part Four has one free-response question worth six points. This question will have multiple parts.

Explaining Your Reasoning

When a free-response question asks to "Justify your answer," "Explain your answer," or "Explain how you determined your answer," the grader is expecting a few clearly written sentences. For these, you don't want to write too little since the grader needs to see that you understand why you did the different steps you did to solve the equation. You also don't want to write too much because if anything you write is not accurate, points can be deducted.

Here is an example followed by two solutions. The first would not get full credit, but the second would.

Example 1

Use algebra to solve for x in the equation $\frac{2}{3}x + 1 = 11$. Justify your steps.

Solution 1 (part credit):

$\frac{2}{3}x + 1 = 11$ $\underline{\phantom{\frac{2}{3}x+1}-1 = -1}$ $\frac{2}{3}x = 10$ $x = 15$	I used algebra to get the x by itself. The answer was $x = 15$.

Solution 2 (full credit):

$\frac{2}{3}x + 1 = 11$ $\underline{\phantom{\frac{2}{3}x+1}-1 = -1}$ $\frac{2}{3}x = 10$ $\frac{3}{2} \cdot \frac{2}{3}x = \frac{3}{2} \cdot 10$ $1x = 15$ $x = 15$	I used the subtraction property of equality to eliminate the $+1$ from the left-hand side. Then to make it so the x had a 1 in front of it, I used the multiplication property of equality and multiplied both sides of the equation by the reciprocal of $\frac{2}{3}$, which is $\frac{3}{2}$. Then since $1 \cdot x = x$, the left-hand side of the equation just became x and the right-hand side became 15.

Computational Errors vs. Conceptual Errors

In the Part Three and Part Four questions, the graders are instructed to take off one point for a "computational error" but half credit for a "conceptual error." This is the difference between these two types of errors.

If a four point question was $x - 1 = 2$ and a student did it like this,

$$x - 1 = 2$$
$$\underline{+1 = +1}$$
$$x = 4$$

the student would lose one point out of 4 because there was one computational error since $2 + 1 = 3$ and not 4.

Had the student done it like this,

$$x - 1 = 2$$
$$\underline{-1 = -1}$$
$$x = 1$$

the student would lose half credit, or 2 points, since this error was conceptual. The student thought that to eliminate the -1, he should subtract 1 from both sides of the equation.

Either error might just be careless, but the conceptual error is the one that gets the harsher deduction.

12.3   KNOW YOUR CALCULATOR

Which Calculator Should You Use?

The two calculators used for this book are the TI-84 and the TI-Nspire. Both are very powerful. The TI-84 is somewhat easier to use for the functions needed for this test. The TI-Nspire has more features for courses in the future. The choice is up to you. This author prefers the TI-84 for the Algebra II Regents.

Graphing calculators come with manuals that are as thick as the book you are holding. There are also plenty of video tutorials online for learning how to use advanced features of the calculator. To become an expert user, watch the online tutorials or read the manual.

Clearing the Memory

You may be asked at the beginning of the test to clear the memory of your calculator. When practicing for the test, you should clear the memory too so you are practicing under test-taking conditions.

This is how you clear the memory.

For the TI-84:

Press [2ND] and then [+] to get to the MEMORY menu. Then press [7] for Reset.

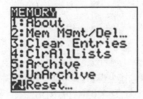

Use the arrows to go to [ALL] for All Memory. Then press [1].

Press [2] for Reset.

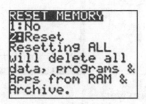

The calculator will be reset as if brand new condition. The one setting that you may need to change is to turn the diagnostics on if you need to calculate the correlation coefficient.

For the TI-Nspire:

The TI-Nspire must be set to Press-To-Test mode when taking the Algebra Regents. Turn the calculator off by pressing [ctrl] and [home]. Press and hold [esc] and then press [home].

While in Press-to-Test mode, certain features will be deactivated. A small green light will blink on the calculator so a proctor can verify the calculator is in Press-to-Test mode.

To exit Press-to-Test mode, use a USB cable to connect the calculator to another TI-Nspire. Then from the home screen on the calculator in Press-to-Test mode, press [doc], [9] and select Exit Press-to-Test.

Use Parentheses

The calculator always uses the order of operations where multiplication and division happen before addition and subtraction. Sometimes, though, you may want the calculator to do the operations in a different order.

Suppose at the end of a quadratic equation, you have to round $x = \dfrac{-1+\sqrt{5}}{2}$ to the nearest hundredth. If you enter (–) (1) (+) (2ND) (x^2) (5) (/) (2), it displays

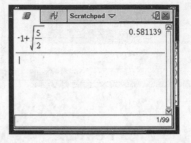

which is not the correct answer.

One reason is that for the TI-84 there needs to be a closing parentheses (or on the TI-Nspire, press [right arrow] to move out from under the radical sign) after the 5 in the square root symbol. Without it, it calculated $-1 + \sqrt{\dfrac{5}{2}}$.

More needs to be done, though, since

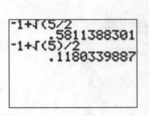

still is not correct. This is the solution to $-1 + \sqrt{\dfrac{5}{2}}$.

To get this correct, there also needs to be parentheses around the entire numerator, $-1 + \sqrt{5}$.

This is the correct answer.

On the TI-Nspire, fractions like this can also be done with [templates].

Using the ANS Feature

The last number calculated with the calculator is stored in something called the ANS variable. This ANS variable will appear if you start an expression with a $+$, $-$, $\times$, or $\div$. When an answer has a lot of digits in it, this saves time and is also more accurate.

If for some step in a problem you need to calculate the decimal equivalent of $\frac{1}{7}$, it will look like this on the TI-84:

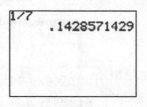

For the TI-Nspire, if you try the same thing, it leaves the answer as $\frac{1}{7}$.

To get the decimal approximation, press [ctrl] and [enter] instead of just [enter].

Now if you have to multiply this by 3, just press [×], and the calculator will display "Ans*"; press [3] and [enter].

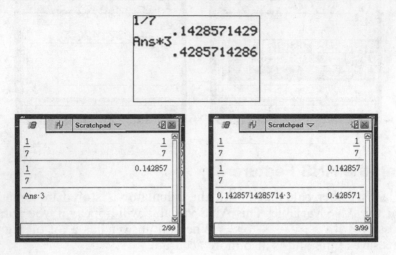

The ANS variable can also help you do calculations in stages. To calculate $x = \dfrac{-1+\sqrt{5}}{2}$ without using so many parentheses as before, it can be done by first calculating $-1 + \sqrt{5}$ and then pressing [÷] and [2] and Ans will appear automatically.

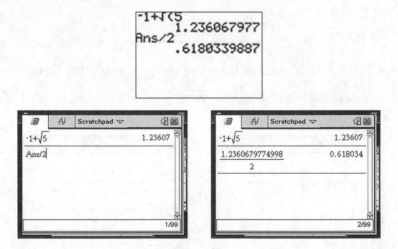

The ANS variable can also be accessed by pressing [2ND] and [–] at the bottom right of the calculator. If after calculating the decimal equivalent of $\dfrac{1}{7}$ you wanted to subtract $\dfrac{1}{7}$ from 5, for the TI-84 press [5], [–], [2ND], [ANS], and [ENTER].

For the TI-Nspire press [5], [–], [ctrl], [ans], and [enter].

12.4 USE THE REFERENCE SHEET

In the back of the Algebra Regents booklet is a reference sheet with 17 conversion facts, like inches to centimeters and quarts to pints, and also 17 formulas. Many of these conversion facts and formulas will not be needed for an individual test, but the quadratic formula and the arithmetic sequence formula are the two that will come in the handiest.

High School Math Reference Sheet

1 inch = 2.54 centimeters	1 kilometer = 0.62 mile	1 cup = 8 fluid ounces
1 meter = 39.37 inches	1 pound = 16 ounces	1 pint = 2 cups
1 mile = 5280 feet	1 pound = 0.454 kilogram	1 quart = 2 pints
1 mile = 1760 yards	1 kilogram = 2.2 pounds	1 gallon = 4 quarts
1 mile = 1.609 kilometers	1 ton = 2000 pounds	1 gallon = 3.785 liters
		1 liter = 0.264 gallon
		1 liter = 1000 cubic centimeters

Triangle	$A = \frac{1}{2}bh$	Pythagorean Theorem	$a^2 + b^2 = c^2$
Parallelogram	$A = bh$	Quadratic Formula	$x = \dfrac{-b \pm \sqrt{b^2 - 4ac}}{2a}$
Circle	$A = \pi r^2$	Arithmetic Sequence	$a_n = a_1 + (n - 1)d$
Circle	$C = \pi d$ or $C = 2\pi r$	Geometric Sequence	$a_n = a_1 r^{n-1}$
General Prisms	$V = Bh$	Geometric Series	$S_n = \dfrac{a_1 - a_1 r^n}{1 - r}$ where $r \neq 1$
Cylinder	$V = \pi r^2 h$	Radians	$1 \text{ radian} = \dfrac{180}{\pi}$ degrees
Sphere	$V = \frac{4}{3}\pi r^3$	Degrees	$1 \text{ degree} = \dfrac{\pi}{180}$ radians
Cone	$V = \frac{1}{3}\pi r^2 h$	Exponential Growth/Decay	$A = A_0 e^{k(t - t_0)} + B_0$
Pyramid	$V = \frac{1}{3}Bh$		

12.5 HOW MANY POINTS DO YOU NEED TO PASS?

The Algebra II Regents exam is scored out of a possible 86 points. Unlike most tests given in the year by your teacher, the score is not then turned into a percent out of 86. Instead each test has a conversion sheet that varies from year to year. For the June 2016 test, the conversion sheet looked like this.

Raw Score	Scale Score	Raw Score	Scale Score	Raw Score	Scale Score	Raw Score	Scale Score
88	100	65	83	42	61	19	33
87	99	64	82	41	60	18	31
86	99	63	81	40	59	17	30
85	98	62	80	39	58	16	28
84	97	61	80	38	56	15	27
83	97	60	79	37	55	14	25
82	96	59	78	36	54	13	24
81	95	58	77	35	53	12	22
80	95	57	76	34	52	11	21
79	94	56	75	33	51	10	19
78	93	55	74	32	49	9	18
77	92	54	73	31	48	8	16
76	92	53	72	30	47	7	14
75	91	52	71	29	46	6	13
74	90	51	70	28	44	5	11
73	89	50	69	27	43	4	9
72	88	49	68	26	42	3	7
71	88	48	67	25	41	2	5
70	87	47	66	24	39	1	3
69	86	46	65	23	38	0	0
68	86	45	64	22	37		
67	85	44	63	21	35		
66	84	43	62	20	34		

On this test, 30 points became a 65, 57 points became a 75, and 73 points became an 85. This means that for this examination a student who got 30 out of 86, which is just 35% of the possible points, would get a 65 on this exam. 57 out of 86 is 66%, but this scaled to a 75. 73 out of 86, however, is actually 85% and became an 85. So in the past there has been a curve on the exam for lower scores, though the scaling is not released until after the exam.

Answers and Solution Hints to Practice Exercises

CHAPTER 1

Section 1.1

A

1. (1)   4. (1)   7. (1)   10. (4)
2. (2)   5. (4)   8. (2)   11. (3)
3. (4)   6. (4)   9. (1)

B

1. $10x^2 - 15x + 20$
2. Zahra did not distribute the $-$ through. The correct answer is $3x^2 + 4x + 6$.
3. $a = 2$
4. $4x^2 + 10x + 25$ is a perfect square trinomial because $\left(\dfrac{10}{2}\right)^2 = 25$. It factors to $\left(x + \dfrac{10}{2}\right)^2 = (x+5)^2$. So $a = 5$.
5. $x^3 + 9x^2 + 27x + 27$

Section 1.2

A

1. (1)   3. (3)   5. (4)   7. (2)   9. (1)
2. (1)   4. (4)   6. (2)   8. (1)   10. (3)

B

1. $(2x + 1)(x + 3)$
2. $(x + 17)(x + 39)$
3. 81
4. $29 \cdot 19$, because $24^2 - 5^2 = (24 + 5)(24 - 5)$
5. $(x^2 - 9)(x^2 - 4) = (x - 3)(x + 3)(x - 2)(x + 2)$

Section 1.3

A

1. (1)   3. (2)   5. (3)   7. (3)   9. (4)
2. (2)   4. (4)   6. (4)   8. (1)   10. (3)

B

1. $4^3 + (5)(4^2) - (7)(4) + 3 = 119$
2. 7   The remainder equals the value of the function at $x = 3$.
3. 100   The second polynomial is 6 greater than the first polynomial. Thus, the remainder of the second polynomial must be $6 + 94$.
4. Jose checked to see if $7^3 - 12 \cdot 7^2 + 43 \cdot 7 - 56 = 0$.
5. $(x + 3)$

Section 1.4

A

1. (2)	3. (4)	5. (1)	7. (3)	9. (3)
2. (1)	4. (4)	6. (1)	8. (4)	10. (2)

B

1. Both are right and can complete the question to get $x = \pm 5$.
2. This factors by grouping to become $(x + 4)(x^2 - 9) = 0$, which then becomes $(x + 4)(x - 3)(x + 3) = 0$. The solutions are -4, 3, and -3.
3. The solutions are $(x - 2)(x - 4)(x - 5) = 0$ or $x^3 - 11x^2 + 38x - 40 = 0$.
4. $x = \dfrac{-b \pm \sqrt{b^2 + 16}}{2}$
5. Delilah noticed that $x^2 + 10x + 16$ factors to $(x + 2)(x + 8)$.

Section 1.5

A

1. (4)	3. (2)	5. (1)	7. (2)	9. (3)
2. (1)	4. (1)	6. (1)	8. (3)	10. (1)

B

1.

2. 8

3. The x-coordinate of the vertex is 5. The y-coordinate is $5^2 - 10(5) + 23 = -2$. In vertex form, the equation is $y = (x - 5)^2 - 2$.

4. The vertex is halfway between the focus and directrix. So the vertex is at $(4, 2)$. The equation is $y = a(x - 4)^2 + 2$. Since $p = \dfrac{3-1}{2} = 1$, $a = \dfrac{1}{4p} = \dfrac{1}{4 \cdot 1} = \dfrac{1}{4}$. So the complete equation is $y = \dfrac{1}{4}(x-4)^2 + 2$.

5. $\overline{PF} = \overline{PD}$ because any point on the parabola is equidistant from the focus and the directrix so $\overline{PF} = \overline{PD} = 8$.

Section 1.6

A

1. (1)   3. (3)   5. (4)   7. (1)   9. (4)
2. (1)   4. (3)   6. (4)   8. (2)   10. (2)

B
1. Intercepts at $(2, 0)$, $(-2, 0)$, $(-3, 0)$, $(0, -2)$

2. $y = \dfrac{1}{3}(x-3)^2(x+1)$
3. $y = a(x - 2)^2(x + 1)$ where a is any real number
4. The a-value of the first graph is positive. The a-value of the second graph is negative.
5. $(-5, 0)$ and $(-1, 0)$. Divide by $(x - 2)$, and factor the quotient $x^2 + 6x + 5$ into $(x + 5)(x + 1)$.

CHAPTER 2

Section 2.1

A

1. (1)   3. (1)   5. (4)   7. (2)   9. (2)
2. (3)   4. (3)   6. (4)   8. (1)   10. (3)

B

1. Braylon is right. $\dfrac{2x + 5}{2} = x + \dfrac{5}{2}$

2. $\dfrac{1}{x} - \dfrac{1}{x + 1} = \dfrac{x + 1}{x(x + 1)} - \dfrac{x}{x(x + 1)} = \dfrac{1}{x(x + 1)}$

3. $\left(\dfrac{x}{x(x + h)} - \dfrac{x + h}{x(x + h)} \right) \div h = \dfrac{-h}{x(x + h)} \cdot \dfrac{1}{h} = -\dfrac{1}{x(x + h)}$

4. Talia forgot to distribute the minus sign. It should be $5x - 15 - 2x - 4$ in the numerator to get $\dfrac{3x - 19}{(x + 2)(x - 3)}$.

5. $\dfrac{2x^2 + 4x + 1}{(x + 1)(x + 2)}$

6. $\dfrac{(x + 4)(x + 5)}{(x - 3)}$. Since the denominator has the factors $(x - 3)$ and $(x + 3)$, the numerator must be divisible by one of these factors. The numerator factors to $(x + 3)(x + 4)(x + 5)$.

Section 2.2

A

1. (3)   3. (3)   5. (3)   7. (1)   9. (4)
2. (4)   4. (3)   6. (2)   8. (1)   10. (1)

B

1. Originally, 6 people pay $4 each. Then 8 people pay $3 each.
2. Paxton is correct. The first and third terms are undefined when $x = 2$.
3. They drove 60 mph. The equation $\dfrac{240}{x} + \dfrac{4}{5} = \dfrac{240}{x - 10}$, where x = rate from New York to Boston.
4. The equation simplifies to $2x^2 + 11x - 30 = 0$. So $x = -7.5$ or $x = 2$.
5. The equation simplifies to $7x + 5 = 26$, so $x = 3$.

Section 2.3

A

1. (2)   2. (1)   3. (3)   4. (1)   5. (2)

CHAPTER 3

Section 3.1

A

1. (2)   3. (3)   5. (3)   7. (2)   9. (3)
2. (2)   4. (4)   6. (1)   8. (3)   10. (2)

B

1. Ashlynn is right because $5^3 \cdot 5^4 = 5^{3 + 4} = 5^7$.
2. $5.2 \times 10^{-4} = 5.2 \times \dfrac{1}{10^4} = 5.2 \times \dfrac{1}{10,000} = 0.00052$.

3. $2^{11} = 2^{10} \cdot 2^1 = 1{,}024 \cdot 2 = 2{,}048.$

4. $9^{1.5} = 9^{\frac{3}{2}} = \left(\sqrt[2]{9}\right)^3 = 3^3 = 27$

5. $\dfrac{1}{64} \cdot 125 \cdot \dfrac{1}{9} = \dfrac{125}{576}$

Section 3.2

A

1. (1)   3. (4)   5. (3)   7. (4)   9. (4)
2. (2)   4. (1)   6. (2)   8. (4)   10. (3)

B

1. They intersect at (2.7, 20).

2. They are both right. $2^{-x} = (2^{-1})^x = \left(\dfrac{1}{2}\right)^x.$

3. The equation simplifies to $-2x + 6 = -6x + 18$. Thus, $x = 3$.
4. Between 3 and 4 since $4^3 = 64$ and $4^4 = 256$.
5. $3 \cdot 2^x$ is not equivalent to 6^x.

Section 3.3

A

1. (3)   3. (3)   5. (3)   7. (3)   9. (1)
2. (4)   4. (1)   6. (4)   8. (1)   10. (2)

B

1. $\log_3 1{,}000$ is greater. Since 3 is less than 4, it would have to be raised to a larger power to become 1,000.
2. As an exponential equation, it would become $7^x = 7^{123456789}$. So $x = 123456789$.
3. Subtract 7 from 262, and then divide by 3. You get $5^x = 85$. So $x = \log_5 85 \approx 2.76$.

4. They are both correct: $\dfrac{\log 263}{\log 7} = \dfrac{\ln 263}{\ln 7} = \log_7 263.$

5.

Section 3.5

A

1. (3) 3. (1) 5. (2) 7. (2) 9. (2)
2. (1) 4. (4) 6. (3) 8. (4) 10. (1)

B

1. a) 61 million b) $T = 37$, year 2037
2. a) 57 degrees b) 4 minutes
3. $A = 200(1.01)^{12T}$
4. $100 \cdot {}^e(0.49^2) = 110.296$ Reagan will have \$110.25, and David will have \$110.30. So David will have more money.
5. Since $240,000 = \dfrac{1}{1,267,200} \cdot 2^N$, $N = 41$ folds.

CHAPTER 4

Section 4.1

A

1. (2) 3. (1) 5. (1) 7. (2) 9. (2)
2. (4) 4. (2) 6. (4) 8. (3) 10. (1)

B

1. Mia is correct. There is no distributive property for radicals.
2. No. $\sqrt{49} + \sqrt{576} = 7 + 24 = 31$
3. $\sqrt{8^2 - 4^2} = \sqrt{64 - 16} = \sqrt{48} = 4\sqrt{3}$
4. $\dfrac{6\sqrt{3}}{3} = 2\sqrt{3}$
5. $25 - 10\sqrt{3} + 10\sqrt{3} - 4 \cdot 3 = 13$

Section 4.2

A

1. (1)   3. (1)   5. (4)   7. (2)   9. (1)
2. (4)   4. (1)   6. (2)   8. (1)   10. (3)

B

1.

The absolute value is $\sqrt{(-6)^2+(8)^2} = \sqrt{100} = 10$.

2. $(x^2 + 4)(x^2 + 9) = 0$
   $x^2 = -4$ or $x^2 = -9$
   $x = \pm 2i$ or $x = \pm 3i$

3. $4i$

4. $\dfrac{7}{4} + \dfrac{3}{2}i$

5. $\sqrt{1.9^2 + 0.8^2} \approx 2.06 > 2$

Section 4.3

A

1. (2)   3. (2)   5. (2)   7. (3)   9. (4)
2. (3)   4. (3)   6. (1)   8. (1)   10. (2)

B

1. $x = 13$
2. $x = 4$
3. $x^2 = 9$ has the solution set $\{-3, 3\}$.
4. $\left(\sqrt{x-4}+\sqrt{x+4}\right)^2$ does not equal $x - 4 + x + 4$. Instead, it equals
   $x - 4 + 2\sqrt{x-4}\sqrt{x+4} + x + 4$.
5. $x = 3$ $\sqrt{x + 6} = 2 + \sqrt{4 - x}, x + 6 = 4 + 4\sqrt{4 - x} + 4 - x, 2x - 2 =$
   $4\sqrt{4 - x}, 4x^2 - 8x + 4 = 16(4 - x), 4x^2 + 8x - 60 = 0,$
   $4(x^2 + 2x - 15) = 0, 4(x - 3)(x + 5) = 0.$

Section 4.4

A

1. (3)	3. (1)	5. (2)	7. (1)	9. (3)
2. (2)	4. (4)	6. (1)	8. (1)	10. (1)

B

1.

2. $y = \sqrt{x-2} - 4$

3. They are both right. $\sqrt{4x} = 2\sqrt{x}$.

4. They both go through (0, 0) and (1, 1). $y = \sqrt[3]{x}$ is in quadrants I and III, while $y = \sqrt{x}$ is in only quadrant I.

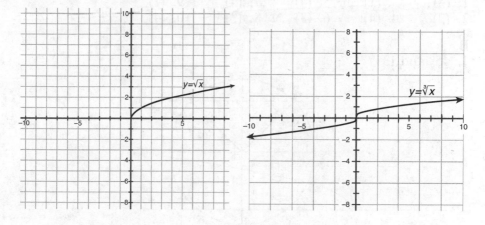

5. $\sqrt{x+2} + 3 = 6$

CHAPTER 5

Section 5.1

A

1. (2)   3. (4)   5. (2)   7. (4)   9. (2)
2. (1)   4. (3)   6. (4)   8. (4)   10. (1)

B

1. a) By the Pythagorean theorem, $AC \approx 0.8$.
   b) By similar triangles, $DF \approx 2.43$.
2. $FG = 8 \cdot \sin 40° \approx 5.14$, $EF = 8 \cdot \cos 40° \approx 6.13$
3. The $\sin \angle AOB$ is the y-coordinate of point B, which is 0.6, $\sin^{-1}(0.6) = 37°$. Therefore, $\angle AOB = 180° - 37°$.
4. $180° - 25° = 155°$, $180° + 25° = 205°$, $360° - 25° = 335°$
5. $\sin^{-1}.4848 \approx 29°$. $\sin \theta < 0$ in quadrants III and IV so $\theta = 180° + 29°$ $= 209°$. $\cos 209° \approx -.8746$

Section 5.2

A

1. (4)	3. (1)	5. (4)	7. (4)	9. (1)
2. (2)	4. (4)	6. (3)	8. (3)	10. (4)

B

1. 45° and 135°

2. 115° and 245°

3. The horizontal line $y = 2$ does not intersect the circle.
4. $x = 180° + \sin^{-1}(0.8192) \approx 235°$ and $x = 360° - \sin^{-1}(0.8192) \approx 305°$
5. When $45° < x < 225°$, the y-coordinate of the point on the unit circle is greater than the x-coordinate.

Section 5.3

A

1. (3)	3. (4)	5. (4)	7. (4)	9. (2)
2. (3)	4. (3)	6. (1)	8. (1)	10. (4)

B

1. One radian is smaller than 60°. It is approximately 57.3°. Also triangle AOB cannot be equilateral since $\overline{OA} = \overline{OB} = \overparen{AB} > \overline{AB}$.

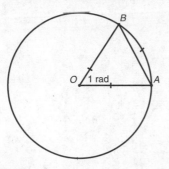

2. $\dfrac{5\pi}{3}$ radians = 300° and $\sin 300° = -\dfrac{\sqrt{3}}{2} \approx -0.8660$.

3. 1 degree, 1 radian, 1 right angle

4. $\angle AOB$ is bigger. Since BC is sin $\angle AOB$ and arc $\overparen{BA}$ is the same as $\angle AOB$ in radians, $\angle AOB = \overparen{BA} > \overline{BC} = \sin\angle AOB$.

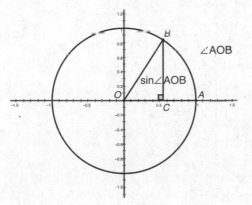

5. 100 gradians. $\dfrac{\pi}{2}$ radians is 90°, which is $\dfrac{1}{4}$ of a circle. So $\dfrac{1}{4}$ of 400 gradians is 100 gradians.

Section 5.4

A

1. (3)	3. (2)	5. (3)	7. (1)	9. (1)
2. (2)	4. (4)	6. (4)	8. (4)	10. (4)

B

1.

2.

3. $y = -3\sin(2x) + 4$

4. $y = 4\cos\left(\dfrac{\pi}{30}x\right)$

5.

Section 5.5

A

1. (4)   3. (1)   5. (2)   7. (1)   9. (1)
2. (1)   4. (3)   6. (2)   8. (3)   10. (4)

Section 5.6

B

1. a) 30 seconds
   b) 44 feet
   c)

2. a)

   b) $h = -30\cos\left(\dfrac{\pi}{40}t\right) + 35$

   c) 46.5 feet
   d) 23 seconds and 58 seconds

3. a) $y = -30\cos\left(\dfrac{\pi}{6}x\right) + 55$

   b) 40 degrees
   c) $x = 8$ (September) and $x = 7$ (August)

Section 5.7

A

1. (3)   3. (4)   5. (1)   7. (4)   9. (2)
2. (2)   4. (1)   6. (1)   8. (4)   10. (2)

B

1. $\sin\theta\tan\theta + \cos\theta = \sin\theta\,\dfrac{\sin\theta}{\cos\theta} + \dfrac{\cos^2\theta}{\cos\theta} = \dfrac{\sin^2\theta + \cos^2\theta}{\cos\theta} = \dfrac{1}{\cos\theta} = \sec\theta$

2. $\dfrac{\cos\theta}{1+\sin\theta} = \dfrac{\cos\theta(1-\sin\theta)}{(1+\sin\theta)(1-\sin\theta)} = \dfrac{\cos\theta(1-\sin\theta)}{\cos^2\theta} = \dfrac{1-\sin\theta}{\cos\theta}$

3. $(\cos\theta - \sin\theta)(\cos\theta + \sin\theta) + 2\sin^2\theta = \cos^2\theta - \sin^2\theta + 2\sin^2\theta$

$$= \cos^2\theta + \sin^2\theta = 1$$

4. $(\cot\theta + \csc\theta)(\cot\theta - \csc\theta) = \left(\dfrac{\cos\theta}{\sin\theta} + \dfrac{1}{\sin\theta}\right)\left(\dfrac{\cos\theta}{\sin\theta} - \dfrac{1}{\sin\theta}\right)$

$$= \dfrac{\cos^2\theta}{\sin^2\theta} - \dfrac{1}{\sin^2\theta} = \dfrac{\cos^2\theta - 1}{\sin^2\theta}$$

$$= \dfrac{-1(1 - \cos^2\theta)}{\sin^2\theta} = \dfrac{-\sin^2\theta}{\sin^2\theta} = -\sin\theta$$

5. $\dfrac{\sin^2\theta + \cos^2\theta}{\sin^2\theta} = \dfrac{1}{\sin^2\theta}$ becomes $1 + \dfrac{\cos^2\theta}{\sin^2\theta} = \csc^2\theta$ becomes

$1 + \cot^2\theta = \csc^2\theta$.

CHAPTER 6

Section 6.1

A

1. (4)	3. (1)	5. (1)	7. (1)	9. (4)
2. (1)	4. (3)	6. (2)	8. (4)	10. (2)

B
1. The substitution method is easier to use if one of the equations has either x or y isolated.
2. Both are correct.
3. (–4, 2)
4. Tickets were bought for 3 adults and 7 children.
5. $m = 15, n = 5$

Section 6.2

B
1. 4
2. (3, 1, 4)
3. (2, 4, 1)
4. (4, –1, 3)
5. (–2, 3, –4)

CHAPTER 7

Section 7.1

A

1. (3)   3. (1)   5. (4)   7. (1)   9. (4)
2. (2)   4. (4)   6. (1)   8. (3)   10. (3)

B

1. Griffin is correct. $g(f(3)) = 17$
2. $3x^2 - 24x + 48$
3. $g(2) = 4, f(g(2)) = f(4) = 5$
4. It is $y = f(x)$ shifted up by 2 units.

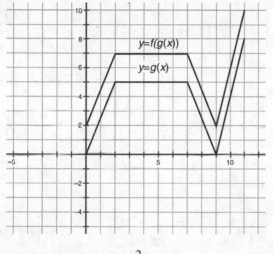

5. $(x-2)^2 = x^2 - 2, x = \dfrac{3}{2}$

Section 7.2

A

1. (2)   3. (1)   5. (3)   7. (2)   9. (4)
2. (4)   4. (1)   6. (1)   8. (3)   10. (3)

B

1. 6

2.

3. $f^{-1}(x) = \dfrac{x-3}{7}$

4. x

5. $f^{-1}(x) = \dfrac{x}{2}, g^{-1}(x) = x - 7$

$$2x + 7 = g(f(x)) = 17$$
$$g^{-1}(g(f(x))) = g^{-1}(17)$$
$$f(x) = 10$$
$$f^{-1}(f(x)) = f^{-1}(10)$$
$$x = 5$$

CHAPTER 8

Section 8.1

A

1. (1)	3. (2)	5. (1)	7. (4)	9. (3)
2. (2)	4. (3)	6. (4)	8. (3)	10. (3)

B

1. $\dfrac{1,000(1-1.05^{21})}{1-1.05} = \$35,719$

2. The 20th term of the second sequence is $2^{19} = 524,288$ while the 20th term of the first sequence is 200,000.

3. 55 The first ten terms are 1, 1, 2, 3, 5, 8, 13, 21, 34, 55.

4. $a_4 = 2,356$

5. $5 \cdot 3^{n-1} = 2,657,205$

$$3^{n-1} = 531,441$$

$$n - 1 = \log_3 531,441$$

$$n - 1 = 12$$

$$n = 13$$

CHAPTER 9

Section 9.1

A

1. (2)   3. (4)   5. (2)   7. (1)   9. (2)

2. (1)   4. (1)   6. (1)   8. (4)   10. (3)

B

1. {1A, 2A, 3A, 4A, 5A, 6A, 1B, 2B, 3B, 4B, 5B, 6B, 1C, 2C, 3C, 4C, 5C, 6C}

2. $\dfrac{3}{8}$

3. 25

4. a) $\dfrac{90}{200}$ b) $\dfrac{50}{90}$ c) $\dfrac{65}{130}$

5. a) {BBBB, BBBG, BBGB, BBGG, BGBB, BGBG, BGGB, BGGG, GBBB, GBBG, GBGB, GBGG, GGBB, GGBG, GGGB, GGGG}

b) There is a $\dfrac{6}{16}$ chance of 2 boys 2 girls, $\dfrac{4}{16}$ chance of 3 boys 1 girl, and a $\dfrac{4}{16}$ chance of 3 girls 1 boy. So there is an $\dfrac{8}{16}$ chance of 3 of one gender and 1 of the other, which is greater than $\dfrac{6}{16}$ for 2 of each gender.

Section 9.2

A

1. (1)   3. (3)   5. (3)   7. (4)   9. (1)
2. (1)   4. (1)   6. (3)   8. (2)   10. (2)

B

1. Various answers. One could be B = "I have an outdoor picnic this Saturday."
2. $.4 \times (1 - .3) = .28$
3. No. P(A or B) = P(A) + P(B) – P(A and B). She did not subtract P(A and B).
4. If events are independent, $P(B|A) = P(B)$ so $P(B|A) = .2$.
5. 0. It is impossible for both to win since they are playing against each other. These are not independent events.

Section 9.3

A

1. (3)   3. (3)   5. (4)   7. (2)   9. (2)
2. (2)   4. (3)   6. (3)   8. (2)   10. (1)

B

1. Various answers are acceptable such as, "She gets lung cancer," "She has yellow teeth," "She has bad breath," etc.
2. For independent events, P(A and B) = P(A) · P(B). If P(A and B) = .14, these are independent events. Otherwise, they are dependent events.
3. These are not independent events, otherwise P(Layla goes to the party given that Charlotte goes to the party) would equal P(Layla goes to the party).
4. There are three ways to verify if two events are independent:
   1) If P(S) = P($S|B$),
   2) If P(B) = P($B|A$), or
   3) If P(A and B) = P(A) · P(B). P(S) = P($S|B$) = .4, P(B) = P($B|S$) = .5, and $P(S$ and $B) = P(S) \cdot P(B) = .20$. By either of the three tests, these are independent events.

5. There are three ways to check for independence:

1) Does P(science) = P(science|straight hair)? No, since $\dfrac{45}{150} \neq \dfrac{10}{60}$.

2) Does P(straight hair) = P(straight hair | science)? No, since $\dfrac{60}{150} \neq \dfrac{10}{45}$.

3) Does P(straight hair and science) = P(straight hair) · P(science)? No, since $\dfrac{10}{150} \neq \dfrac{2700}{22,500}$. None of these equalities work, so these are dependent events.

CHAPTER 10

Section 10.1

A

1. (3)   3. (2)   5. (1)   7. (4)   9. (1)
2. (1)   4. (4)   6. (4)   8. (1)   10. (2)

Section 10.2

A

1. (1)   3. (4)   5. (1)   7. (4)   9. (2)
2. (4)   4. (1)   6. (3)   8. (3)   10. (1)

B

1. a) mean = 70.5, standard deviation = 17.7
   b) 70.5 − 17.7 = 52.8 and 70.5 + 17.7 = 88.2. So of the ten numbers, 8 or 80% are within one standard deviation of the mean. (c) If the set was perfectly normal, 68% would be within one standard deviation of the mean. So this set could be considered approximately normal.
2. The mean is 40, and the standard deviation is 3.9. All of the numbers are between 40 − 2(3.9) = 32.2 and 40 + 2(3.9) = 47.8. So all are within two standard deviations of the mean.
3. 14 people is 7%. If this was a normal distribution, it would be expected for just 2.2% to be 82 inches or more. This suggests that the data are not normally distributed.

4. The mean is 73.65, and the standard deviation is 5.01. To be within two standard deviations of the mean, numbers would need to be between $73.65 - 2(5.01) = 63.6$ and $73.65 + 2(5.01) = 83.7$. The numbers 62 and 84 are not within two standard deviations of the mean. The rest are.

5. 68.3% within one standard deviation of the mean suggest that it could be a normal distribution. However, more information would need to be known to be sure.

CHAPTER 11

Section 11.1

B

1. One way to reduce bias is to choose the subjects randomly from a list rather than choosing people who are in a certain place or at a certain time. The wording of the questions should also avoid language that could interfere with the way the subjects answer the questions.

2. As an experiment, you would somehow have to force one group of students to have a certain amount of sleep. This could be done by having some kind of late meeting or event that causes the group to get less sleep than the other group. As an observation, you would not do anything to affect the amount of sleep that people get and then record what you observe about people who have different amounts of sleep.

3. One way is to assign each student in the grade a random number. Then have a computer generate random numbers to choose the groups without any chance of bias.

4. When the question is worded with language like "boring," it can cause responders to get a negative feeling about the question. Questions should use more neutral language.

5. This is an experiment since the researcher is controlling who is exposed to different temperatures.

Section 11.2

A

1. (1)	3. (1)	5. (1)	7. (4)	9. (3)
2. (4)	4. (2)	6. (1)	8. (2)	10. (2)

Glossary of Algebra II Terms

A

Algebraic identity Two expressions usually involving polynomials that can be shown to be equal to one another.

Amplitude The distance between the middle of a sine or cosine curve to the top of that sine or cosine curve. When the equation is $y = a\sin(bx) + d$ or $y = a\cos(bx) + d$, the amplitude is a.

Arithmetic sequence A sequence of numbers like 3, 7, 11, 15, ... where each number is the same amount more (or less) than the previous number. That difference is called d. In this case, $d = 4$.

Arithmetic series The sum of the terms in an arithmetic sequence. $3 + 7 + 11 + 15$ is an arithmetic series.

Asymptote A line that a curve gets closer and closer to without ever reaching.

B

Base In an exponential expression, the base is the number being raised to a power. In the expression $5 \cdot 2^x$, 2 is the base. In a logarithmic expression, the base is the small subscript number after the word "log." In the expression $\log_3 81 = 4$, the base is 3.

Binomial A polynomial with two terms linked by either a plus (+) or minus (−) sign. The expression $2x + 5$ is a binomial.

C

Coefficient The number that gets multiplied by a variable in a mathematical expression. In the expression $5x^2$, the coefficient of the x^2-term is 5.

Complex number A number of the form $a + bi$ where a and b are real numbers. $5 + 3i$ is a complex number.

Composite function When the expression that defines one function is put into another function. If $f(x) = 3x$ and $g(x) = x^2$, then $g(f(x)) = (3x)^2$ is a composite function.

Compound interest When money in a bank gets interest on an initial deposit and then later gets more interest based on the original deposit and on the interest already earned.

Constant A number. The term in a polynomial that has no variable part. In the polynomial $2x + 3$, the constant is 3.

Cosine The ratio between the adjacent side and the hypotenuse of a right triangle. Also the x-coordinate of a point on the unit circle.

Cubic equation An equation of the form $ax^3 + bx^2 + cx + d = 0$.

Cubic function A function of the form $f(x) = ax^3 + bx^2 + cx + d$.

D

Degree The highest exponent in a polynomial. The degree is $x^3 + 5x^2 + 7x + 2$ is 3.

Degree A unit of angle measurement that is $\frac{1}{360}$ of a circle.

Dependent events When the probability of one event is affected by whether or not some other event happened, the two events are called dependent events. The event "The baseball game will be canceled" and the event "It will rain" are dependent events.

Difference of perfect squares A way of factoring a polynomial like $x^2 - 3^2$ into $(x - 3)(x + 3)$. In general, an expression of the form $a^2 - b^2$ can be factored into $(a - b)(a + b)$.

Directrix Every point on a parabola is equidistant from a point and a line. That line is called the directrix.

E

e A mathematical constant that is approximately 2.72. It is involved in many questions related to modeling growth.

Elimination method Multiplying one or more equations in a system of equations by a constant and then adding the equations in a way that makes one of the variables cancel out. The system $x + 2y = 5$ and $3x - 2y = 7$ becomes $4x = 12$ when the two equations are added together.

Even function A function whose graph is symmetric with respect to the *y*-axis.

Experimental study When a researcher has one group of subjects do one thing and another group do another thing and then studies how the two groups react.

An example is giving half of the subjects one type of medicine and the rest of them another type of medicine.

Explicit sequence formula A way of describing a sequence based on relating the position of a term to the number in that position. $a_n = 5n + 8$ is an explicit sequence formula.

Exponential equation An equation that involves an exponential expression. The variable is in the exponent. $3 \cdot 2^x = 96$ is an exponential equation.

Exponential expression An expression that has a variable as an exponent. $3 \cdot 2^x$ is an exponential expression.

Exponential function A function that has a variable as an exponent. $f(x) = 3 \cdot 2^x$ is an exponential function.

F

Factor by grouping A way of factoring that involves separating a polynomial into two or more groups, finding a common factor in each group, and, if possible, factoring a common factor out of what remains. The expression $x^3 + 3x^2 + 2x + 6 = x^2(x + 3) + 2(x + 3) = (x + 3)(x^2 + 2)$ is an example of factoring by grouping.

Factoring Finding two factors whose product is a given number or polynomial. $15 = 3 \cdot 5$ is a way of factoring 15. $x^2 - 5x + 6 = (x - 2)(x - 3)$ is a way of factoring $x^2 - 5x + 6$.

Factors When a number or polynomial is divided by a number or polynomial and there is no remainder, the divisor is a factor of the dividend. Since $\frac{15}{3} = 5$ with remainder 0, 3 is a factor of 15.

Factor theorem A theorem that states that if the remainder when you divide $P(x)$ by $x - a$ is 0, then $x - a$ is a factor of $P(x)$. For example, if you divide $P(x) = x^2 + 5x + 6$ by $x + 2$, you get $x + 3$ with remainder 0, so $x + 2$ is a factor of $P(x)$.

Finite arithmetic series formula The formula $S_n = \frac{(a_1 + a_n)n}{2}$ for summing the terms of a finite arithmetic series.

Finite geometric series formula The formula $S_n = \frac{a_1(1 - r^n)}{1 - r}$ for summing the terms of a finite geometric series.

Focus Every point on a parabola is equidistant from a point and a line. That point is called the focus of the parabola.

FOIL A way to remember the four multiplications when multiplying a binomial by another binomial. When multiplying $(2x + 3)(4x + 5)$, F (for firsts) represents $2x \cdot 4x$, O (for outers) represents $2x \cdot 5$, I (for inners) represents $3 \cdot 4x$, and L (for lasts) represents $3 \cdot 5$.

Frequency The number of sine or cosine curves that can fit into a 360-degree interval.

Function A rule that takes a number or an expression as an input and converts it into a number or an expression in a predicable way. The function $f(x) = x^2$ takes a number or expression and turns it into the

number or expression squared, for example $f(3) = 3^2 = 9$.

G

Geometric sequence A sequence of numbers like 3, 6, 12, 24, ... where each number is equal to the previous number in the sequence multiplied by the same number. That number, the common ratio, is called r. In this case, $r = 2$.

Geometric series The sum of the terms in a geometric sequence. $3 + 6 + 12 + 24$ is a geometric series.

Given When calculating the probability of B given A, the given means that event A has already happened and may, or may not, affect the probability that event B will also happen.

Greatest common factor Of all the common factors between two numbers or polynomials, the ones that are largest. The numbers 12 and 18 have the common factors 1, 2, 3, and 6, so 6 is the greatest common factor between 12 and 18.

Growth rate In the exponential expression $a(1 + r)^x$, the r is the growth rate. For example, in $100(1 + 0.07)^x$, the growth rate is 0.07.

H

Horizontal shift When all the points of a graph move right or left the same number of units, it is called a horizontal shift.

Horizontal shrink When the x-coordinates of every point in a graph are divided by the same number, it is called a horizontal shrink.

I

i   A shorthand way of writing $\sqrt{-1}$.

Imaginary number   A number that is a multiple of the imaginary unit *i*. $5i$ is an imaginary number.

Independent events   When the probability of one event is not affected by whether or not some other event happened, the two events are called independent events. The event "I wake up at 7:00 A.M." and "It will rain today" are independent events.

Inverse function   If a function $f(x)$ has an inverse $f^{-1}(x)$, then $f^{-1}(f(x)) = x$. For example, if $f(x) = x + 1$, $f^{-1}(x) = x - 1$. An inverse function undoes whatever the original function did.

L

Like terms   When two monomials with the same variable part are added or subtracted, it is called combining like terms. In the expression $2x^2 + 5x + 3x^2 + 7$, the $2x^2$ and the $3x^2$ are like terms and can be combined to form $5x^2$.

Linear function   A function whose largest exponent is 1. $f(x) = 3x + 7$ is a linear function.

ln *x*   A shorthand for writing $\log_e x$.

$\log_b x$ The number that *b* would have to be raised to in order to get a result of *x*. For example, $\log_5 125 = 3$ since $5^3 = 125$.

Logarithm   An exponent that tells what a base needs to be raised to in order to become another number. The logarithm of 25 base 5 or $\log_5 25 = 2$ since $5^2 = 25$.

M

Monomial   A polynomial that has just one term. The polynomial $3x^5$ is a monomial.

N

Normal distribution   When a set of data points resembles a bell curve with 68% being within one standard deviation, 95% being within two standard deviations, it is known as a normal distribution.

O

Observational study   When a statistical researcher collects data by studying the behavior of the subjects without interfering with their environment.

Odd function   A function whose graph is symmetric with respect to the origin.

P

Perfect square trinomial   A polynomial of the form $x^2 + 2ax + a^2$ or $x^2 - 2ax + a^2$ that can be factored into $(x + a)^2$ or $(x - a)^2$. For example, $x^2 + 6x + 9$ can be factored into $(x + 3)^2$.

Period   The number of degrees (or radians) in one cycle of a sine or cosine curve.

Polynomial   An expression like $2x^3 + 5x^2 - 3x + 7$ that consists of one or more terms, each term having a coefficient and a variable raised to a power.

Population mean   The average of an entire population for some characteristic. For example, if out of 10,000 people studied their average height is 64 inches, 64 is the population mean.

Population proportion   The percent of an entire population that has some characteristic. For example, if out of 10,000 people studied 52% are boys, 0.52 is the population proportion.

Probability   The chance of something happening, ranging from 10 (impossible) to 1 (certain).

Q

Quadratic equation   An equation of the form $ax^2 + bx + c = 0$.

Quadratic formula   The formula
$$x = \frac{-b \pm \sqrt{b^2 - 4ac}}{2a}$$ used to find the solutions to the equation $ax^2 + bx + c = 0$.

Quadratic function   A function whose largest exponent is 2. $f(x) = x^2 - 3x + 7$ is a quadratic function.

R

Radian   A unit of angle measurement that is equal to approximately 57 degrees.

Radical equation   An equation involving a radical expression. $\sqrt{x+2} = 5$ is a radical equation.

Radical expression   An expression involving a radical symbol. $\sqrt{x+2}$ is a radical expression.

Randomization test   When a computer is used to simulate an ideal experiement with the goal of seeing if something that already happened was likely to happen or not.

Rational equation   An equation that involves one or more rational expressions. $\frac{3}{x-2} + \frac{4}{x-2} = 5$ is a rational equation.

Rational expression   An expression that has a fraction with a polynomial of degree at least one in the denominator. $\frac{3}{x+1}$ is a rational expression.

Rational Function   A function that has a fraction with a polynomial of degree at least one in the denominator. $f(x) = \frac{3}{x+1}$ is a rational function.

Recursive sequence formula   A way of describing a sequence based on how each term relates to the preceding term. $a_1 = 3$, $a_n = 5 + a_{n-1}$ is a recursive sequence formula.

Remainder theorem   A theorem that states that when you divide a polynomial $P(x)$ by $x - a$, the remainder will always equal $P(a)$. For example, if you divide the polynomial $P(x) = x^2 + 5x + 6$ by $x - 3$, you will get a remainder of $P(3) = 30$.

S

Sample mean   The average of a sample that has some characteristic. For example, if out of 10,000 people 100 people are randomly chosen and of those 100 people their average height is 61 inches, then 61 is the sample mean for that sample.

Sample proportion   The percent of a sample that has some characteristic. For example, if out of 10,000 people 100 people are randomly chosen and of those 100 people 38% are boys, then 0.38 is the sample proportion.

Sample space   A list of all the possible outcomes of a probability experiment.

Sampling distribution of the sample mean When multiple groups of samples are taken from a larger population and the average of some attribute of each sample group is calculated and graphed in a histogram.

Sampling distribution of the sample proportion When multiple groups of samples are taken from a larger population and the percent of some attribute of each sample group is calculated and graphed in a histogram.

Simplified radical form When the number inside a square root (or cube root) sign has no factors that are perfect squares (or perfect cubes). The expression $\sqrt{18}$ is not in simplified radical form since $18 = 2 \cdot 9$. $3\sqrt{2}$ is equal to $\sqrt{18}$ but is in simplified radical form.

Sine The ratio between the opposite side and the hypotenuse of a right triangle. Also the y-coordinate of a point on the unit circle.

Standard deviation A way of measuring how spread out a data set is. The smallest standard deviation is 0. It occurs when all the elements of a set are equal.

Standard form of a quadratic equation An equation is the form $y = ax^2 + bx + c$.

Statistical significance When the data from an experiment are randomly mixed together multiple times with the goal of seeing if a difference between two groups is likely to happen.

Substitution method A system of equations where one of the variables is solved in terms of the other variable and then substituted into the other equation. For example, in the system of equations $y = 2x + 1$, $3x + 2y = 10$, the y in the second equation can be replaced by $2x + 1$ to become $3x + 2(2x + 1) = 10$.

Survey A question or set of questions answered by a subject in a statistical study.

System of equations Two or more equations that have two or more variables. A system of equations generally has one solution that is an ordered pair (or triple) with numbers that satisfy each of the equations.

T

Tangent The ratio between the opposite side and the adjacent side of a right triangle.

Trigonometric function A function that involves a trigonometry ratio, like sine, cosine, or tangent. $f(x) = 3\sin(x) + 1$ is a trigonometric function.

Trigonometric identity Two expressions involving trigonometric ratios that can be shown to be equal to one another.

Trinomial A polynomial with three terms linked by plus (+) and/or minus (−) signs. The expression $x^2 + 5x + 6$ is a trinomial.

Two-way frequency table A chart used for listing the different ways that things can either have or not have certain attributes.

U

Unit circle A circle with a radius of 1 and center at (0, 0).

V

Variable A letter that is used to represent a number. Often this letter represents the thing to be solved for in an equation. In the expression $2x + 5$, the variable is x.

Venn diagram A way of organizing data that has two or more intersecting circles surrounded by a rectangle.

Vertex form of a quadratic equation An equation of the form $y = a(x - h)^2 + k$. When graphed, it is a parabola with vertex (h, k).

Vertical shift When all the points of a graph move up or down the same number of units, it is called a vertical shift.

Vertical stretch When the y-coordinates of every point in a graph are multiplied by the same number, it is called a vertical stretch.

Z

z-score A measure of how many standard deviations above (positive z-score) or below (negative z-score) a number is from the mean.

THE ALGEBRA II REGENTS EXAMINATION

The Regents Examination in Algebra II is a three-hour exam that is divided into four parts with a total of 37 questions. All 37 questions must be answered. Part I consists entirely of regular multiple-choice questions. Parts II, III, and IV each contain a set of questions that must be answered directly in the question booklet. You are required to show how you arrived at the answers for the questions in Parts II, III, and IV. The acompanying table shows how the exam breaks down.

Question Type	Number of Questions	Credit Value
Part I: Multiple choice	24	$24 \times 2 = 48$
Part II: 2-credit open ended	8	$8 \times 2 = 16$
Part III: 4-credit open ended	4	$4 \times 4 = 16$
Part IV: 6-credit open ended	1	$1 \times 6 = 6$
	Total = 37 questions	Total = 86 points

How Is the Exam Scored?

- Each of your answers to the 24 multiple-choice questions in Part I will be scored as either right or wrong.
- Solutions to questions in Part II, III, and IV that are not completely correct may receive partial credit according to a special rating guide that is provided by the New York State Education Department. In order to receive full credit for a correct answer to a question in Parts II, III, or IV, you must show or explain how you arrived at your answer by indicating the key steps taken, including appropriate formula substitutions, diagrams, graphs, and charts. A correct numerical answer with no work shown will receive only 1 credit.
- The raw scores for the four parts of the test are added together. The maximum total raw score for the Algebra II Regents Examination is 86 points. Using a special conversion chart that is provided by the New York State Education Department, your total raw score will be equated to a final test score that falls within the usual 0 to 100 scale.

What Type of Calculator Is Required?

Graphing calculators are *required* for the Algebra II Regents Examination. During the administration of the Regents exam, schools are required to make a graphing calculator available for the exclusive use of each student. You will need to use your calculator to work with trigonometric functions of angles, find roots of numbers, and perform routine calculations.

Knowing how to use a graphing calculator gives you more options when deciding how to solve a problem. Rather than solving a problem algebraically with pen and paper, it may be easier to solve the same problem using a graph or table created by a graphing calculator. A graphical or numerical solution using a calculator can also be used to help confirm an answer obtained by solving the problem algebraically.

Are Any Formulas Provided?

The Algebra II Regents Examination test booklet will include a reference sheet containing the formulas in the accompanying table. Keep in mind that you may be required to know other formulas that are not included in this sheet. Please see page 373 for these formulas.

What Else Should I Know?

- Do not omit any questions from Part I. Since there is no penalty for guessing, make certain that you record an answer for each of the 24 multiple-choice questions.
- If the method of solution is not stated in the problem, choose an appropriate method (numerical, graphical, or algebraic) with which you are most comfortable.
- If you solve a problem in Parts II, III, or IV using a trial-and-error approach, show the work for at least three guesses with appropriate checks. Should the correct answer be reached on the first trial, you must further illustrate your method by showing that guesses below and above the correct guess do not work.
- Avoid rounding errors when using a calculator. Unless otherwise directed, the (pi) key on a calculator should be used in computations involving the constant π rather than the common rational approximation of 3.14 or $\frac{22}{7}$. When performing a sequence of calculations in which the result of one calculation is used in a second calculation, do not round off. Instead, use the full power/display of the calculator by performing a "chain" calculation, saving intermediate results in the calculator's memory. Unless otherwise specified, rounding, if required, should be done only when the *final* answer is reached.
- Check that each answer is in the requested form. If a specific form is not required, answers may be left in any equivalent form, such as $\sqrt{75}$, $5\sqrt{3}$, or 8.660254038 (the full power/display of the calculator).
- If a problem requires using a formula that is not provided in the question, check the formula reference sheet in the test booklet to see if it is listed. Clearly write any formula you use before making any appropriate substitutions. Then evaluate the formula in step-by-step fashion.
- For any problem solved in Parts II, III, and IV using a graphing calculator, you must indicate how the calculator was used to obtain the answer such as by copying graphs or tables created by your calculator together with the equations used to produce them. When copying graphs, label each graph with its equation, state the dimensions of the viewing window, and indentify the intercepts and any points of intersection with their coordinates. Whenever appropriate, indicate the rationale of your approach.

Examination
June 2018
Algebra II

HIGH SCHOOL MATH REFERENCE SHEET

Conversions

1 inch = 2.54 centimeters	1 cup = 8 fluid ounces
1 meter = 39.37 inches	1 pint = 2 cups
1 mile = 5280 feet	1 quart = 2 pints
1 mile = 1760 yards	1 gallon = 4 quarts
1 mile = 1.609 kilometers	1 gallon = 3.785 liters
	1 liter = 0.264 gallon
1 kilometer = 0.62 mile	1 liter = 1000 cubic centimeters
1 pound = 16 ounces	
1 pound = 0.454 kilogram	
1 kilogram = 2.2 pounds	
1 ton = 2000 pounds	

Formulas

Triangle	$A = \dfrac{1}{2}bh$
Parallelogram	$A = bh$
Circle	$A = \pi r^2$
Circle	$C = \pi d$ or $C = 2\pi r$

Formulas (continued)

General Prisms $\qquad V = Bh$

Cylinder $\qquad V = \pi r^2 h$

Sphere $\qquad V = \dfrac{4}{3}\pi r^3$

Cone $\qquad V = \dfrac{1}{3}\pi r^2 h$

Pyramid $\qquad V = \dfrac{1}{3}Bh$

Pythagorean Theorem $\qquad a^2 + b^2 = c^2$

Quadratic Formula $\qquad x = \dfrac{-b \pm \sqrt{b^2 - 4ac}}{2a}$

Arithmetic Sequence $\qquad a_n = a_1 + (n-1)d$

Geometric Sequence $\qquad a_n = a_1 r^{n-1}$

Geometric Series $\qquad S_n = \dfrac{a_1 - a_1 r^n}{1-r}$ where $r \neq 1$

Radians $\qquad 1 \text{ radian} = \dfrac{180}{\pi} \text{ degrees}$

Degrees $\qquad 1 \text{ degree} = \dfrac{\pi}{180} \text{ radians}$

Exponential Growth/Decay $\qquad A = A_0 e^{k(t - t_0)} + B_0$

PART I

Answer all 24 questions in this part. Each correct answer will receive 2 credits. No partial credit will be allowed. For each statement or question, write in the space provided the numeral preceding the word or expression that best completes the statement or answers the question. [48 credits]

1 The graphs of the equations $y = x^2 + 4x - 1$ and $y + 3 = x$ are drawn on the same set of axes. One solution of this system is

(1) $(-5, -2)$ (3) $(1, 4)$

(2) $(-1, -4)$ (4) $(-2, -1)$ 1 _____

2 Which statement is true about the graph of $f(x) = \left(\dfrac{1}{8}\right)^x$?

(1) The graph is always increasing.

(2) The graph is always decreasing.

(3) The graph passes through $(1, 0)$.

(4) The graph has an asymptote, $x = 0$.   2 _____

3 For all values of x for which the expression is defined, $\dfrac{x^3 + 2x^2 - 9x - 18}{x^3 - x^2 - 6x}$, in simplest form, is equivalent to

(1) 3 (3) $\dfrac{x + 3}{x}$

(2) $-\dfrac{17}{2}$ (4) $\dfrac{x^2 - 9}{x(x - 3)}$ 3 _____

4 A scatterplot showing the weight, w, in grams, of each crystal after growing t hours is shown below.

The relationship between weight, w, and time, t, is best modeled by

(1) $w = 4^t + 5$ (3) $w = 5(2.1)^t$

(2) $w = (1.4)^t + 2$ (4) $w = 8(.75)^t$ 4 ____

5 Where i is the imaginary unit, the expression $(x + 3i)^2 - (2x - 3i)^2$ is equivalent to

(1) $-3x^2$ (3) $-3x^2 + 18xi$

(2) $-3x^2 - 18$ (4) $-3x^2 - 6xi - 18$ 5 ____

6 Which function is even?

(1) $f(x) = \sin x$ (3) $f(x) = |x - 2| + 5$

(2) $f(x) = x^2 - 4$ (4) $f(x) = x^4 + 3x^3 + 4$ 6 ____

7 The function $N(t) = 100e^{-0.023t}$ models the number of grams in a sample of cesium-137 that remain after t years. On which interval is the sample's average rate of decay the fastest?

(1) $[1, 10]$ (3) $[15, 25]$

(2) $[10, 20]$ (4) $[1, 30]$ 7 _____

8 Which expression can be rewritten as $(x + 7)(x - 1)$?

(1) $(x + 3)^2 - 16$

(2) $(x + 3)^2 - 10(x + 3) - 2(x + 3) + 20$

(3) $\dfrac{(x - 1)(x^2 - 6x - 7)}{(x + 1)}$

(4) $\dfrac{(x + 7)(x^2 + 4x + 3)}{(x + 3)}$ 8 _____

9 What is the solution set of the equation $\dfrac{2}{x} - \dfrac{3x}{x + 3} = \dfrac{x}{x + 3}$?

(1) $\{3\}$ (3) $\{-2, 3\}$

(2) $\left\{\dfrac{3}{2}\right\}$ (4) $\left\{-1, \dfrac{3}{2}\right\}$ 9 _____

10 The depth of the water at a marker 20 feet from the shore in a bay is depicted in the graph below.

If the depth, d, is measured in feet and time, t, is measured in hours since midnight, what is an equation for the depth of the water at the marker?

(1) $d = 5 \cos\left(\dfrac{\pi}{6}t\right) + 9$ (3) $d = 9 \sin\left(\dfrac{\pi}{6}t\right) + 5$

(2) $d = 9 \cos\left(\dfrac{\pi}{6}t\right) + 5$ (4) $d = 5 \sin\left(\dfrac{\pi}{6}t\right) + 9$ 10 _____

11 On a given school day, the probability that Nick oversleeps is 48% and the probability he has a pop quiz is 25%. Assuming these two events are independent, what is the probability that Nick oversleeps and has a pop quiz on the same day?

(1) 73%   (3) 23%

(2) 36%   (4) 12%   11 _____

12 If $x - 1$ is a factor of $x^3 - kx^2 + 2x$, what is the value of k?

(1) 0   (3) 3

(2) 2   (4) −3   12 _____

13 The profit function, $p(x)$, for a company is the cost function, $c(x)$, subtracted from the revenue function, $r(x)$. The profit function for the Acme Corporation is $p(x) = -0.5x^2 + 250x - 300$ and the revenue function is $r(x) = -0.3x^2 + 150x$. The cost function for the Acme Corporation is

(1) $c(x) = 0.2x^2 - 100x + 300$
(2) $c(x) = 0.2x^2 + 100x + 300$
(3) $c(x) = -0.2x^2 + 100x - 300$
(4) $c(x) = -0.8x^2 + 400x - 300$

13 _____

14 The populations of two small towns at the beginning of 2018 and their annual population growth rate are shown in the table below.

Town	Population	Annual Population Growth Rate
Jonesville	1240	6% increase
Williamstown	890	11% increase

Assuming the trend continues, approximately how many years after the beginning of 2018 will it take for the populations to be equal?

(1) 7   (3) 68
(2) 20   (4) 125

14 _____

15 What is the inverse of $f(x) = x^3 - 2$?

(1) $f^{-1}(x) = \sqrt[3]{x} + 2$ (3) $f^{-1}(x) = \sqrt[3]{x + 2}$
(2) $f^{-1}(x) = \pm\sqrt[3]{x} + 2$ (4) $f^{-1}(x) = \pm\sqrt[3]{x + 2}$

15 _____

16 A 4th degree polynomial has zeros -5, 3, i, and $-i$.
    Which graph could represent the function defined by
    this polynomial?

(1)   (3)

(2)   (4)   16 _____

17 The weights of bags of Graseck's Chocolate Candies are normally distributed with a mean of 4.3 ounces and a standard deviation of 0.05 ounces. What is the probability that a bag of these chocolate candies weighs less than 4.27 ounces?

(1) 0.2257   (3) 0.7257

(2) 0.2743   (4) 0.7757   17 _____

18 The half-life of iodine-131 is 8 days. The percent of the isotope left in the body d days after being introduced is $I = 100\left(\dfrac{1}{2}\right)^{\frac{d}{8}}$. When this equation is written in terms of the number e, the base of the natural logarithm, it is equivalent to $I = 100e^{kd}$. What is the approximate value of the constant, k?

(1) −0.087   (3) −11.542

(2) 0.087   (4) 11.542   18 __

19 The graph of $y = \log_2 x$ is translated to the right 1 unit and down 1 unit. The coordinates of the x-intercept of the translated graph are

(1) $(0, 0)$ (3) $(2, 0)$

(2) $(1, 0)$ (4) $(3, 0)$ 19 _____

20 For positive values of x, which expression is equivalent to $\sqrt{16x^2} \bullet x^{\frac{2}{3}} + \sqrt[3]{8x^5}$?

(1) $6\sqrt[5]{x^3}$ (3) $4\sqrt[3]{x^2} + 2\sqrt[3]{x^5}$

(2) $6\sqrt[3]{x^5}$ (4) $4\sqrt{x^3} + 2\sqrt[5]{x^3}$ 20 _____

21 Which equation represents a parabola with a focus of
$(-2, 5)$ and a directrix of $y = 9$?

(1) $(y - 7)^2 = 8(x + 2)$
(2) $(y - 7)^2 = -8(x + 2)$
(3) $(x + 2)^2 = 8(y - 7)$
(4) $(x + 2)^2 = -8(y - 7)$

21 _____

22 Given the following polynomials

$$x = (a + b + c)^2$$
$$y = a^2 + b^2 + c^2$$
$$z = ab + bc + ac$$

Which identity is true?

(1) $x = y - z$ (3) $x = y - 2z$
(2) $x = y + z$ (4) $x = y + 2z$

22 _____

23 On average, college seniors graduating in 2012 could
compute their growing student loan debt using the
function $D(t) = 29{,}400(1.068)^t$, where t is time in years.
Which expression is equivalent to $29{,}400(1.068)^t$ and
could be used by students to identify an approximate
daily interest rate on their loans?

(1) $29{,}400\left(1.068^{\frac{1}{365}}\right)^t$ (3) $29{,}400\left(1 + \dfrac{0.068}{365}\right)^t$

(2) $29{,}400\left(\dfrac{1.068}{365}\right)^{365t}$ (4) $29{,}400\left(1.068^{\frac{1}{365}}\right)^{365t}$

23 _____

24 A manufacturing plant produces two different-sized containers of peanuts. One container weighs x ounces and the other weighs y pounds. If a gift set can hold one of each size container, which expression represents the number of gift sets needed to hold 124 ounces?

(1) $\dfrac{124}{16x + y}$

(3) $\dfrac{124}{x + 16y}$

(2) $\dfrac{x + 16y}{124}$

(4) $\dfrac{16x + y}{124}$

24 _____

PART II

Answer all 8 questions in this part. Each correct answer will receive 2 credits. Clearly indicate the necessary steps, including appropriate formula substitutions, diagrams, graphs, charts, etc. For all questions in this part, a correct numerical answer with no work shown will receive only 1 credit. [16 credits]

25 A survey about television-viewing preferences was given to randomly selected freshmen and seniors at Fairport High School. The results are shown in the table below.

Favorite Type of Program

	Sports	Reality Show	Comedy Series
Senior	83	110	67
Freshman	119	103	54

A student response is selected at random from the results. State the *exact* probability the student response is from a freshman, given the student prefers to watch reality shows on television.

26 On the grid below, graph the function
$f(x) = x^3 - 6x^2 + 9x + 6$ on the domain $-1 \le x \le 4$.

27 Solve the equation $2x^2 + 5x + 8 = 0$. Express the answer in $a + bi$ form.

28 Chuck's Trucking Company has decided to initiate an Employee of the Month program. To determine the recipient, they put the following sign on the back of each truck.

How's My Driving?

Call 1-555-DRIVING

The driver who receives the highest number of positive comments will win the recognition. Explain *one* statistical bias in this data collection method.

29 Determine the quotient and remainder when
$(6a^3 + 11a^2 - 4a - 9)$ is divided by $(3a - 2)$.

Express your answer in the form $q(a) + \dfrac{r(a)}{d(a)}$.

30 The recursive formula to describe a sequence is shown below.

$$a_1 = 3$$
$$a_n = 1 + 2a_{n-1}$$

State the first four terms of this sequence.

Can this sequence be represented using an explicit geometric formula? Justify your answer.

31 The Wells family is looking to purchase a home in a suburb of Rochester with a 30-year mortgage that has an annual interest rate of 3.6%. The house the family wants to purchase is $152,500 and they will make a $15,250 down payment and borrow the remainder. Use the formula below to determine their monthly payment, to the *nearest dollar*.

$$M = \frac{P\left(\dfrac{r}{12}\right)\left(1 + \dfrac{r}{12}\right)^n}{\left(1 + \dfrac{r}{12}\right)^n - 1}$$

M = monthly payment
P = amount borrowed
r = annual interest rate
n = total number of monthly payments

32 An angle, θ, is in standard position and its terminal side passes through the point $(2, -1)$. Find the *exact* value of $\sin \theta$.

PART III

Answer all 4 questions in this part. Each correct answer will receive 4 credits. Clearly indicate the necessary steps, including appropriate formula substitutions, diagrams, graphs, charts, etc. For all questions in this part, a correct numerical answer with no work shown will receive only 1 credit. [16 credits]

33 Solve algebraically for all values of x:

$$\sqrt{6-2x} + x = 2(x+15) - 9$$

34 Joseph was curious to determine if scent improves memory. A test was created where better memory is indicated by higher test scores. A controlled experiment was performed where one group was given the test on scented paper and the other group was given the test on unscented paper. The summary statistics from the experiment are given below.

	Scented Paper	Unscented Paper
$\bar{x}$	23	18
s_x	2.898	2.408

Calculate the difference in means in the experimental test grades (scented – unscented).

A simulation was conducted in which the subjects' scores were rerandomized into two groups 1000 times. The differences of the group means were calculated each time. The results are shown below.

Question 34 is continued on the next page.

Question 34 continued.

Use the simulation results to determine the interval represent
ing the middle 95% of the difference in means, to the *nearest
hundredth*.

Is the difference in means in Joseph's experiment statistically
significant based on the simulation? Explain.

35 Carla wants to start a college fund for her daughter Lila. She puts $63,000 into an account that grows at a rate of 2.55% per year, compounded monthly. Write a function, $C(t)$, that represents the amount of money in the account t years after the account is opened, given that no more money is deposited into or withdrawn from the account.

Calculate algebraically the number of years it will take for the account to reach $100,000, to the *nearest hundredth of a year*.

36 The height, $h(t)$ in cm, of a piston, is given by the equation $h(t) = 12 \cos\left(\dfrac{\pi}{3}t\right) + 8$, where t represents the number of seconds since the measurements began.

Determine the average rate of change, in cm/sec, of the piston's height on the interval $1 \le t \le 2$.

At what value(s) of t, to the *nearest tenth of a second*, does $h(t) = 0$ in the interval $1 \le t \le 5$? Justify your answer.

PART IV

Answer the question in this part. A correct answer will receive 6 credits. Clearly indicate the necessary steps, including appropriate formula substitutions, diagrams, graphs, charts, etc. A correct numerical answer with no work shown will receive only 1 credit. [6 credits]

37 Website popularity ratings are often determined using models that incorporate the number of visits per week a website receives. One model for ranking websites is $P(x) = \log(x - 4)$, where x is the number of visits per week in thousands and $P(x)$ is the website's popularity rating.

According to this model, if a website is visited 16,000 times in one week, what is its popularity rating, rounded to the *nearest tenth*?

Graph $y = P(x)$ on the axes below.

Question 37 is continued on the next page.

Question 37 continued.

An alternative rating model is represented by $R(x) = \frac{1}{2}x - 6$, where x is the number of visits per week in thousands. Graph $R(x)$ on the same set of axes. For what number of weekly visits will the two models provide the same rating?

Answers
June 2018
Algebra II

Answer Key

PART I

1. (2)	**5.** (3)	**9.** (4)	**13.** (1)	**17.** (2)	**21.** (4)
2. (2)	**6.** (2)	**10.** (4)	**14.** (1)	**18.** (1)	**22.** (4)
3. (3)	**7.** (1)	**11.** (4)	**15.** (3)	**19.** (4)	**23.** (4)
4. (2)	**8.** (1)	**12.** (3)	**16.** (2)	**20.** (2)	**24.** (3)

PART II

25. $\dfrac{103}{213}$

26.

27. $-\dfrac{5}{4} \pm \dfrac{\sqrt{39}}{4}i$

28. Not a random sampling

29. $2a^2 + 5a + 2 - \dfrac{5}{3a - 2}$

30. 3, 7, 15, 31; this is not a geometric series.

31. $624

32. $\dfrac{-1}{\sqrt{5}}$

PART III

33. −15

34. 5; between −3.07 and 3.13. It is significant because 5 > 3.13.

35. $C(t) = 63{,}000\left(1 + \dfrac{0.0255}{12}\right)^{12t}$;

18.14

36. −12; zeros at 2.2 and 3.8

PART IV

37. 1.1; 14,000 visits

In **Parts II–IV**, you are required to show how you arrived at your answers. For sample methods of solutions, see Barron's *Regents Exams and Answers* for Algebra II.

Examination
June 2019
Algebra II

HIGH SCHOOL MATH REFERENCE SHEET

Conversions

1 inch = 2.54 centimeters

1 meter = 39.37 inches

1 mile = 5280 feet

1 mile = 1760 yards

1 mile = 1.609 kilometers

1 kilometer = 0.62 mile

1 pound = 16 ounces

1 pound = 0.454 kilogram

1 kilogram = 2.2 pounds

1 ton = 2000 pounds

1 cup = 8 fluid ounces

1 pint = 2 cups

1 quart = 2 pints

1 gallon = 4 quarts

1 gallon = 3.785 liters

1 liter = 0.264 gallon

1 liter = 1000 cubic centimeters

Formulas

Triangle	$A = \frac{1}{2}bh$
Parallelogram	$A = bh$
Circle	$A = \pi r^2$
Circle	$C = \pi d$ or $C = 2\pi r$

Formulas (continued)

General Prisms	$V = Bh$
Cylinder	$V = \pi r^2 h$
Sphere	$V = \frac{4}{3}\pi r^3$
Cone	$V = \frac{1}{3}\pi r^2 h$
Pyramid	$V = \frac{1}{3}Bh$
Pythagorean Theorem	$a^2 + b^2 = c^2$
Quadratic Formula	$x = \dfrac{-b \pm \sqrt{b^2 - 4ac}}{2a}$
Arithmetic Sequence	$a_n = a_1 + (n-1)d$
Geometric Sequence	$a_n = a_1 r^{n-1}$
Geometric Series	$S_n = \dfrac{a_1 - a_1 r^n}{1 - r}$ where $r \neq 1$
Radians	1 radian $= \dfrac{180}{\pi}$ degrees
Degrees	1 degree $= \dfrac{\pi}{180}$ radians
Exponential Growth/Decay	$A = A_0 e^{k(t - t_0)} + B_0$

PART I

Answer all 24 questions in this part. Each correct answer will receive 2 credits. No partial credit will be allowed. For each statement or question, write in the space provided the numeral preceding the word or expression that best completes the statement or answers the question. [48 credits]

1 A sociologist reviews randomly selected surveillance videos from a public park over a period of several years and records the amount of time people spent on a smartphone. The statistical procedure the sociologist used is called

(1) a census   (3) an observational study

(2) an experiment   (4) a sample survey   1 _____

2 Which statement(s) are true for all real numbers?

$$\text{I } (x - y)^2 = x^2 + y^2$$
$$\text{II } (x + y)^3 = x^3 + 3xy + y^3$$

(1) I, only   (3) I and II

(2) II, only   (4) neither I nor II   2 _____

3 What is the solution set of the following system of equations?

$$y = 3x + 6$$
$$y = (x + 4)^2 - 10$$

(1) $\{(-5, -9)\}$ (3) $\{(0, 6),(-5, -9)\}$

(2) $\{(5, 21)\}$ (4) $\{(0, 6),(5, 21)\}$ 3 _____

4 Irma initially ran one mile in over ten minutes. She then began a training program to reduce her one-mile time. She recorded her one-mile time once a week for twelve consecutive weeks, as modeled in the graph below.

Number of Weeks

Which statement regarding Irma's one-mile training program is correct?

(1) Her one-mile speed increased as the number of weeks increased.

(2) Her one-mile speed decreased as the number of weeks increased.

(3) If the trend continues, she will run under a six-minute mile by week thirteen.

(4) She reduced her one-mile time the most between weeks ten and twelve.

4 _____

5 A 7-year lease for office space states that the annual rent is $85,000 for the first year and will increase by 6% each additional year of the lease. What will the total rent expense be for the entire 7-year lease?

(1) $42,809.63

(2) $90,425.53

(3) $595,000.00

(4) $713,476.20

5 _____

6 The graph of $y = f(x)$ is shown below.

Which expression defines $f(x)$?

(1) $2x$ (3) $5\left(2^{\frac{x}{2}}\right)$

(2) $5(2^x)$ (4) $5\left(2^{2x}\right)$ 6 _____

7 Given $P(x) = x^3 - 3x^2 - 2x + 4$, which statement is true?

(1) $(x - 1)$ is a factor because $P(-1) = 2$.
(2) $(x + 1)$ is a factor because $P(-1) = 2$.
(3) $(x + 1)$ is a factor because $P(1) = 0$.
(4) $(x - 1)$ is a factor because $P(1) = 0$.   7 _____

8 For $x \geq 0$, which equation is *false*?

(1) $\left(x^{\frac{3}{2}}\right)^2 = \sqrt[4]{x^3}$ (3) $\left(x^{\frac{3}{2}}\right)^{\frac{1}{2}} = \sqrt[4]{x^3}$

(2) $\left(x^3\right)^{\frac{1}{4}} = \sqrt[4]{x^3}$ (4) $\left(x^{\frac{2}{3}}\right)^2 = \sqrt[3]{x^4}$ 8 _____

9 What is the inverse of the function $y = 4x + 5$?

(1) $x = \dfrac{1}{4}y - \dfrac{5}{4}$ (3) $y = 4x - 5$

(2) $y = \dfrac{1}{4}x - \dfrac{5}{4}$ (4) $y = \dfrac{1}{4x + 5}$ 9 _____

10 Which situation could be modeled using a geometric sequence?

(1) A cell phone company charges $30.00 per month for 2 gigabytes of data and $12.50 for each additional gigabyte of data.
(2) The temperature in your car is 79°. You lower the temperature of your air conditioning by 2° every 3 minutes in order to find a comfortable temperature.
(3) David's parents have set a limit of 50 minutes per week that he may play online games during the school year. However, they will increase his time by 5% per week for the next ten weeks.
(4) Sarah has $100.00 in her piggy bank and saves an additional $15.00 each week.   10 _____

11 The completely factored form of
$n^4 - 9n^2 + 4n^3 - 36n - 12n^2 + 108$ is

(1) $(n^2 - 9)(n + 6)(n - 2)$
(2) $(n + 3)(n - 3)(n + 6)(n - 2)$
(3) $(n - 3)(n - 3)(n + 6)(n - 2)$
(4) $(n + 3)(n - 3)(n - 6)(n + 2)$ 11 _____

12 What is the solution when the equation $wx^2 + w = 0$ is solved for x, where w is a positive integer?

(1) -1 (3) 6
(2) 0   (4) $\pm i$ 12 _____

13 A group of students was trying to determine the proportion of candies in a bag that are blue. The company claims that 24% of candies in bags are blue. A simulation was run 100 times with a sample size of 50, based on the premise that 24% of the candies are blue. The approximately normal results of the simulation are shown in the dot plot below.

The simulation results in a mean of 0.254 and a standard deviation of 0.060. Based on this simulation, what is a plausible interval containing the middle 95% of the data?

(1) (0.194, 0.314)   (3) (−0.448, 0.568)
(2) (0.134, 0.374)   (4) (0.254, 0.374)   13 _____

14 Selected values for the functions f and g are shown in the tables below.

x	f(x)
−3.12	−4.88
0	−6
1.23	−4.77
8.52	2.53
9.01	3.01

x	g(x)
−2.01	−1.01
0	0.58
8.52	2.53
13.11	3.01
16.52	3.29

A solution to the equation $f(x) = g(x)$ is

(1) 0   (3) 3.01
(2) 2.53   (4) 8.52   14 _____

15 The expression $6 - (3x - 2i)^2$ is equivalent to

(1) $-9x^2 + 12xi + 10$ (3) $-9x^2 + 10$

(2) $9x^2 - 12xi + 2$ (4) $-9x^2 + 12xi - 4i + 6$ 15 _____

16 A number, minus twenty times its reciprocal, equals eight. The number is

(1) 10 or –2   (3) –10 or –2

(2) 10 or 2   (4) –10 or 2   16 _____

17 A savings account, S, has an initial value of $50. The account grows at a 2% interest rate compounded n times per year, t, according to the function below.

$$S(t) = 50\left(1 + \frac{.02}{n}\right)^{nt}$$

Which statement about the account is correct?

(1) As the value of n increases, the amount of interest per year decreases.

(2) As the value of n increases, the value of the account approaches the function $S(t) = 50e^{0.02t}$.

(3) As the value of n decreases to one, the amount of interest per year increases.

(4) As the value of n decreases to one, the value of the account approaches the function $S(t) = 50(1 - 0.02)^t$.   17 _____

18 There are 400 students in the senior class at Oak Creek High School. All of these students took the SAT. The distribution of their SAT scores is approximately normal. The number of students who scored within 2 standard deviations of the mean is approximately

(1) 75   (3) 300

(2) 95   (4) 380   18 _____

19 The solution set for the equation $b = \sqrt{2b^2 - 64}$ is

(1) {−8}   (3) {±8}

(2) {8}   (4) { }   19 _____

20 Which table best represents an exponential relationship?

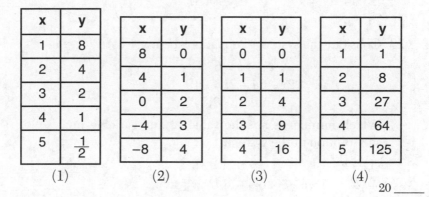

x	y
1	8
2	4
3	2
4	1
5	$\frac{1}{2}$

(1)

x	y
8	0
4	1
0	2
−4	3
−8	4

(2)

x	y
0	0
1	1
2	4
3	9
4	16

(3)

x	y
1	1
2	8
3	27
4	64
5	125

(4)

20 _____

21 A sketch of $r(x)$ is shown below.

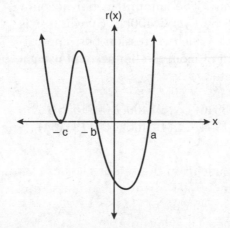

An equation for $r(x)$ could be

(1) $r(x) = (x - a)(x + b)(x + c)$

(2) $r(x) = (x + a)(x - b)(x - c)^2$

(3) $r(x) = (x + a)(x - b)(x - c)$

(4) $r(x) = (x - a)(x + b)(x + c)^2$

21 _____

22 The temperature, in degrees Fahrenheit, in Times Square during a day in August can be predicted by the function $T(x) = 8\sin(0.3x - 3) + 74$, where x is the number of hours after midnight. According to this model, the predicted temperature, to the *nearest degree* Fahrenheit, at 7 P.M. is

(1) 68   (3) 77

(2) 74   (4) 81   22 _____

23 Consider the system of equations below:

$$x + y - z = 6$$
$$2x - 3y + 2z = -19$$
$$-x + 4y - z = 17$$

Which number is *not* the value of any variable in the solution of the system?

(1) −1   (3) 3

(2) 2   (4) −4   23 _____

24 Camryn puts $400 into a savings account that earns 6% annually. The amount in her account can be modeled by $C(t) = 400(1.06)^t$ where t is the time in years. Which expression best approximates the amount of money in her account using a weekly growth rate?

(1) $400(1.001153846)^t$ (3) $400(1.001153846)^{52t}$

(2) $400(1.001121184)^t$ (4) $400(1.001121184)^{52t}$ 24 _____

PART II

Answer all 8 questions in this part. Each correct answer will receive 2 credits. Clearly indicate the necessary steps, including appropriate formula substitutions, diagrams, graphs, charts, etc. For all questions in this part, a correct numerical answer with no work shown will receive only 1 credit. [16 credits]

25 The table below shows the number of hours of daylight on the first day of each month in Rochester, NY.

Month	Hours of Daylight
Jan.	9.4
Feb.	10.6
March	11.9
April	13.9
May	14.7
June	15.4
July	15.1
Aug.	13.9
Sept.	12.5
Oct.	11.1
Nov.	9.7
Dec.	9.0

Given the data, what is the average rate of change in hours of daylight per month from January 1st to April 1st?

Interpret what this means in the context of the problem.

26 Algebraically solve for x:

$$\frac{7}{2x} - \frac{2}{x+1} = \frac{1}{4}$$

27 Graph $f(x) = \log_2(x + 6)$ on the set of axes below.

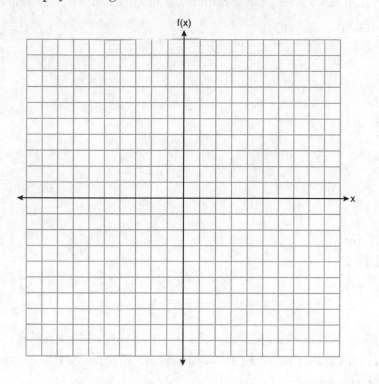

28 Given $\tan \theta = \dfrac{7}{24}$, and θ terminates in Quadrant III, determine the value of $\cos \theta$.

29 Kenzie believes that for $x \geq 0$, the expression $\left(\sqrt[7]{x^2}\right)\left(\sqrt[5]{x^3}\right)$ is equivalent to $\sqrt[35]{x^6}$. Is she correct? Justify your response algebraically.

30 When the function $p(x)$ is divided by $x - 1$ the quotient is $x^2 + 7 + \dfrac{5}{x - 1}$. State $p(x)$ in standard form.

31 Write a recursive formula for the sequence 6, 9, 13.5, 20.25, . . .

32 Robin flips a coin 100 times. It lands heads up 43 times, and she wonders if the coin is unfair. She runs a computer simulation of 750 samples of 100 fair coin flips. The output of the proportion of heads is shown below.

Do the results of the simulation provide strong evidence that Robin's coin is unfair? Explain your answer.

PART III

Answer all 4 questions in this part. Each correct answer will receive 4 credits. Clearly indicate the necessary steps, including appropriate formula substitutions, diagrams, graphs, charts, etc. For all questions in this part, a correct numerical answer with no work shown will receive only 1 credit. [16 credits]

33 Factor completely over the set of integers: $16x^4 - 81$

Sara graphed the polynomial $y = 16x^4 - 81$ and stated "All the roots of $y = 16x^4 - 81$ are real." Is Sara correct? Explain your reasoning.

34 The half-life of a radioactive substance is 15 years.

Write an equation that can be used to determine the amount, $s(t)$, of 200 grams of this substance that remains after t years.

Determine algebraically, to the *nearest year*, how long it will take for $\frac{1}{10}$ of this substance to remain.

35 Determine an equation for the parabola with focus (4, –1) and directrix $y = -5$. (Use of the grid below is optional.)

36 Juan and Filipe practice at the driving range before playing golf. The number of wins and corresponding practice times for each player are shown in the table below.

	Juan Wins	Filipe Wins
Short Practice Time	8	10
Long Practice Time	15	12

Given that the practice time was long, determine the exact probability that Filipe wins the next match.

Determine whether or not the two events "Filipe wins" and "long practice time" are independent. Justify your answer.

PART IV

Answer the question in this part. A correct answer will receive 6 credits. Clearly indicate the necessary steps, including appropriate formula substitutions, diagrams, graphs, charts, etc. A correct numerical answer with no work shown will receive only 1 credit. [6 credits]

37 Griffin is riding his bike down the street in Churchville, N.Y. at a constant speed, when a nail gets caught in one of his tires. The height of the nail above the ground, in inches, can be represented by the trigonometric function $f(t) = -13\cos(0.8\pi t) + 13$, where t represents the time (in seconds) since the nail first became caught in the tire.

Determine the period of $f(t)$.

Interpret what the period represents in this context.

Question 37 is continued on the next page.

Question 37 continued.

On the grid below, graph *at least one* cycle of $f(t)$ that includes the y-intercept of the function.

Does the height of the nail ever reach 30 inches above the ground? Justify your answer.

Answers
June 2019

Algebra II

Answer Key

PART I

1. (3)	**5.** (4)	**9.** (2)	**13.** (2)	**17.** (2)	**21.** (4)
2. (4)	**6.** (3)	**10.** (3)	**14.** (4)	**18.** (4)	**22.** (3)
3. (3)	**7.** (4)	**11.** (2)	**15.** (1)	**19.** (2)	**23.** (2)
4. (1)	**8.** (1)	**12.** (4)	**16.** (1)	**20.** (1)	**24.** (4)

PART II

25. 1.5

26. $\{-2, 7\}$

27.

28. $\dfrac{-24}{25}$

29. She is not correct.

30. $p(x) = x^3 - x^2 + 7x - 2$

31. $\begin{cases} a_1 = 6 \\ a_n = 1.5a_{n-1} \text{ for } n > 1 \end{cases}$

32. It is likely to be a fair coin.

PART III

33. $(2x - 3)(2x + 3)(4x^2 + 9)$

34. $s(t) = 200\left(\dfrac{1}{2}\right)^{\frac{t}{15}}$ and 50

35. $y = \dfrac{1}{8}\left(x - 4\right)^2 - 3$

36. $\dfrac{12}{27}$, not independent

PART IV

37. Period is 2.5. Will not reach 30 since the maximum is 26.

In **PARTS II–IV**, you are required to show how you arrived at your answers. For sample methods of solutions, see *Barron's Regents Exams and Answers* for Algebra II.

Index